Cognitive Psychology

Jackie Andrade and Jon May

Department of Psychology,
University of Sheffield, Sheffield, UK

BIOS Scientific Publishers
Taylor & Francis Group

LONDON AND NEW YORK

© Garland Science/BIOS Scientific Publishers, 2004

First published 2004

A CIP catalogue record for this book is available from the British Library.

ISBN 1 85996 223 8

Garland Science/BIOS Scientific Publishers, 4 Park Square, Milton Park, Abingdon, Oxon OX14 4RN, UK
and 29 West 35th Street, New York, NY 10001–2299, USA
World Wide Web home page: www.bios.co.uk

Garland Science/BIOS Scientific Publishers is a member of the Taylor & Francis Group

Distributed in the USA by
Fulfilment Center
Taylor & Francis
10650 Toebben Drive
Independence, KY 41051, USA
Toll Free Tel.: +1 800 634 7064; E-mail: taylorandfrancis@thomsonlearning.com

Distributed in Canada by
Taylor & Francis
74 Rolark Drive
Scarborough, Ontario M1R 4G2, Canada
Toll Free Tel.: +1 877 226 2237; E-mail: tal_fran@istar.ca

Distributed in the rest of the world by
Thomson Publishing Services
Cheriton House
North Way
Andover, Hampshire SP10 5BE, UK
Tel.: +44 (0)1264 332424; E-mail: salesorder.tandf@thomsonpublishingservices.co.uk

Library of Congress Cataloging-in-Publication Data

Andrade, Jackie, 1964–
    Instant notes in cognitive psychology / Jackie Andrade and Jon May.
        p. cm.
    Includes bibliographical references and index.
    ISBN 1-85996-223-8 (pbk.)
    1. Cognitive psychology—Textbooks. I. May, Jon. II. Title.

 BF201.A53 2004
 153--dc22
     2003018482

Cover image reproduced with permission from Kaniza (1976), Scientific American.
Production Editor: Andrea Bosher
Typest by Phoenix Photosetting, Chatham, Kent, UK
Printed and bound by Biddles Ltd, King's Lynn, Norfolk

CONTENTS

ABBREVIATIONS

2.5D sketch	Marr's postulated 'two-and-a-half-dimensional sketch'	LTM	long-term memory
		m	meter
3D sketch	Marr's 'three-dimensional sketch'	msec	millisecond
AI	artificial intelligence	NP	noun phrase
AIM	affect infusion model	PDP	parallel distributed processing
ANM	associative network model	PET	positron emission tomography
ASL	American Sign Language	PIN	person identity node
cc	cubic centimetres	PL	phonological loop
CCS	central conceptual structure	PSYCOP	a 'natural deduction' model for reasoning, proposed by Rips (1994).
CRT	choice reaction time		
EEG	electroencephalogram		
ELM	elaboration likelihood model	REM	rapid eye movement
ERP	event-related potential	RSVP	rapid serial visual presentation
fMRI	functional magnetic resonance imaging	SAS	supervisory attentional system
		SIU	semantic information unit
FRU	face recognition unit	STM	short-term memory
g	general intelligence	TLC	Teachable Language Comprehender
g_f	fluid ability		
g_c	crystallized ability	TODAM	theory of distributed associative memory
GPC	grapheme–phoneme correspondence		
		V	verb
IAC model	interactive activation and competition model	VOT	voice onset time
		VP	verb phrase
ICS	interacting cognitive subsystems	VSSP	visuo-spatial sketchpad
IQ	intelligence quotient	WAIS	Wechsler Adult Intelligence Scale
LAD	language acquisition device	WISC	Wechsler Intelligence Scale for Children
LGN	lateral geniculate nucleus		
LSD	lysergic acid diethylamide	ZPD	zone of proximal development

PREFACE

Cognitive psychology, the study of human thought processes, is an increasingly broad discipline. Conventional laboratory studies are now supplemented by research using new techniques of brain imaging and computer modeling. The result for students and teachers of the subject has been the need to choose between hefty textbooks that cover all aspects of the field in detail, but make it hard to see the wood for the trees, and shorter books that offer only partial coverage of the field. *Instant Notes in Cognitive Psychology* aims to cover all the key aspects of the subject in a succinct and structured format, enabling the reader to appreciate the important developments in the field and see how new findings can help to improve our understanding of human cognition.

Students new to cognitive psychology often find the topic rather abstract. It tends to be driven not by discoveries but by attempts to explain phenomena that are already known to us, for example how we suddenly hear someone mention our name in a noisy room or why we remember some things for years and forget others immediately. At first encounter, cognitive psychology can seem a way of making familiar things strange. The benefits of studying the subject are a deeper understanding of the complexity and beauty of the processes that enable us to perform mental feats that we take for granted. We have tried to make things easy for the reader new to the field by writing clearly, dividing the subject matter into manageable chunks, and providing signposts to related topics. We have taken particular care with two topics, emotion and consciousness, that are often poorly covered or neglected altogether by other cognitive psychology texts. These are two of the most exciting areas of psychological research and constitute core aspects of human mental life, influencing many aspects of cognition and contributing hugely to our sense of who we are.

We have written this book for undergraduates taking introductory psychology or more specialized cognitive psychology courses, A-level psychology students, postgraduate students or researchers entering psychology from other disciplines, and people studying for the cognitive psychology component of the British Psychological Society qualifying examination. The book can be used as preparatory reading, a revision aid, or a core text supplemented by lectures and further reading.

Reading this book before starting your course will give you an overview of the area, so that specific details can be readily assimilated as you encounter them. For those using this book as a core or supplementary text, we have provided two types of reference. It is important to know who had a good idea, when, and what they did to test it, so in the text we have cited the authors and dates of influential scientific papers and books. Learning names and dates may seem arduous, but it is common in psychology to refer to theories and studies by the names of their authors. These sources are listed in the references section at the end of the book. Reading these original references is beyond the scope of many basic level courses, so we have also included a further reading section that recommends more approachable texts.

To use this book as a revision aid, we recommend that you begin by reading the keynotes at the start of each topic to check if the topic was covered in your course and to help decide if you need to learn more about it. Then read the main text for more detailed information, checking that you understand the key terms highlighted in bold. Follow the links to related topics so you can see how the topic you are revising fits into the rest of cognitive psychology. Integrating information across modules and across different parts of a course is a way of showing that you really understand the topic and gets you good marks in exams. This book is designed to help you do this, with clearly marked links to related topics and sections. After revising a section, set it aside for a while then read the keynotes again to check you have remembered and understood the main points. Section A is intended to give an introduction to cognitive psychology and the methods used to research it. You may find it useful to read it again at the end of your course or revision period, to check that you really understand the nature of the discipline.

Jackie Andrade and Jon May

DEDICATION

To our parents, for their nature and nuture.

A1 COGNITION

Key notes

Cognition	Cognition is the study of the mental processes underlying our ability to perceive the world, remember, talk about and learn from our experiences, and modify our behavior accordingly. It includes functions such as perception, memory, language and thought.
Assumptions about cognition	The mind is a limited capacity information-processing system that behaves in a law-like fashion. Cognition is the product of top-down and bottom-up processes. Top-down processing refers to the influence of knowledge and expectations on functions such as language, perception and memory. Bottom-up processing is processing driven by an external stimulus. Cognitive functions are often assumed to be modular, that is to operate independently of each other.
Philosophical basis	Functionalism views mental events as causal or functional because they serve to transform incoming information into output (different information or behavior). Mind is the 'software' of the brain and can be studied independently of it. For materialists, the mind is the brain and is studied by investigating brain activity directly. Choosing an approach is partly a matter of choosing an appropriate level of explanation for the topic of interest.

Related topics	Methods of investigation (A2)	Issues in consciousness research (K1)

Cognition

In 1967, Ulric Neisser published a book with the title *Cognitive Psychology*. Although scientists had been researching human thought from a cognitive perspective for a couple of decades before this, Neisser's book helped define cognitive psychology as a discipline. Neisser defined cognition as 'all the processes by which the sensory input is transformed, reduced, elaborated, stored, recovered, and used.' Cognitive psychology is the study of the mental processes underlying our ability to perceive the world, to understand and remember our experiences, to communicate with other people, and to control our behavior. It comprises topics such as perception, attention, memory, knowledge, language, problem solving, reasoning and decision making, and aspects of intelligence, emotion and consciousness.

Think about reading this sentence. The rod cells of your retina respond only to light and the cone cells of your retina respond only to light of specific wavelengths. There is no type of cell in your eye that responds selectively to words. We cannot sense words directly. Rather, when you read this paragraph, your brain must transform the pattern of light reaching your retina into symbols representing words. These symbols trigger retrieval of word meanings from memory. To make sense of the text, you must also take into account the syntax or grammar of the sentences. But you are doing many other things too. You may be filtering out background noise and other sensations, such as the texture of the chair you are sitting on. You may be trying to remember if you have already learned anything about cognitive psychology. You may be deciding to take notes and recalling where you left your pen. You may be wondering if you will do well in your exams or worrying that you left the gas on. All this perceiving, understanding, ignoring, deciding, remembering, wondering and worrying is part of cognition.

Assumptions about cognition

Cognitive psychologists share some basic assumptions about the mind:

- The mind is viewed as an **information processing** system through which people interact with the external world. Information processing refers to the manipulation and transformation of symbols or representations of aspects of the external world. It can be represented by box-and-arrow flowcharts. Many cognitive theories have been represented in this way (e.g. Atkinson and Shiffrin's model of memory, shown in *Fig. 1* of Topic D2).
- The mind has resource and structural limitations, that is, it is a limited-capacity processor. Mental processes behave in a systematic, law-like fashion, hence we can investigate them by studying aspects of human behavior such as reaction times and error rates, and we can generalize from studies of small groups of people to humans in general.
- People are active processors of information. They do not respond passively to sensations but rather they use existing knowledge and mental skills to influence those sensations. In other words, our behavior is the result of **bottom-up** and **top-down processing**. Bottom-up processing begins with external input and travels 'up' through the cognitive system, for example, patterns of light hitting the retina are transformed into information about edges and thence into information about objects. Top-down processing refers to the influence of higher-level cognitive elements (goals, intentions, expectations, knowledge etc.) on lower-level processes (see Section B). For an example of the influence of expectation on perception, look at the picture of a duck in *Fig. 1*.

Many psychologists assume that cognitive processes are **modular**. Modules are clusters of processes that function independently from other clusters of processes. Each module processes one particular type of information, for example visual objects or faces. The assumption of modularity means that cognitive psychologists do not have to be experts in all aspects of cognition simultaneously. Rather, they can study in detail one aspect of cognition, for example visual object recognition or face perception. Modularity also underpins the assumption of **localization of function**, that mental processes map onto specific regions of the brain. Therefore, studying the effects of damage to specific brain regions can tell us about normal cognitive functions (see cognitive neuropsychology in Topic A4). However, modularity can also emerge from the activity of distributed, rather than localized, networks of neurons. This means that modular cognitive functions can be mimicked by connectionist networks (see cognitive science in Topic A4; also Topic E4).

Philosophical basis

A philosophical approach called **functionalism** underpins cognitive psychology. It assumes that mental processes and states are functional, that is, they

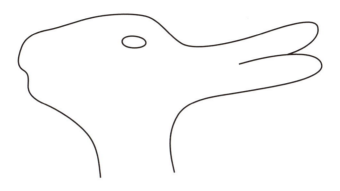

Fig. 1. People can see this ambiguous figure as a duck or a rabbit. The expectation that it was a figure of a duck should have made you see it first as a duck rather than a rabbit, and may make it difficult to 'reverse' your perception to see it as a rabbit.

cause input information to be transformed into output information or behavior. Functionalists propose that the relationship between the mind and brain is analogous to the relationship between the hardware and software of a computer. The most productive way to study human thought and behavior is to study the 'software' of the mind. An alternative approach is **materialism**, the view that the mind and brain are identical therefore human thought and behavior can be understood by studying neural activity in the brain. The two views are not mutually exclusive. Indeed, cognitive psychologists almost invariably hold both. However, whether you are a functionalist or a materialist at heart determines the approach you will prefer for investigating the mind. Functionalist approaches focus on explaining human behavior in terms of information processing and mental functions, whereas materialist approaches focus on reducing mental processes and states to their underlying neural anatomy and biochemistry. Which approach to choose is partly a question of deciding the appropriate **level of explanation** for the phenomenon of interest. For example, the Mediterranean can be described in terms of its geography and tourist industry or in terms of the chemistry of the weak hydrogen bonds that give water its unique properties. One description is better for explaining its economy, the other for explaining its wetness. Memory can be explained in terms of processes such as rehearsal and retrieval (see Section D), or in terms of neurotransmitter release and changes in synaptic sensitivity. The first, cognitive, level of explanation is better for understanding how best to revise for a psychology exam, the second helps explain how the apparently homogenous squishy mass that is your brain can learn and adapt.

Ultimately, we need both types of explanation, functional and biological. With this in mind, Marr (1982) proposed a framework for characterizing cognitive theories. The highest level of explanation, the **computational level**, requires specification of the functions of the cognitive phenomenon under investigation, for example the goals of its component processes and the context in which they operate. The **algorithmic level** requires specification of how the computations or functions can be carried out, for example how the input is represented and the details of how it is transformed into output. The lowest, **hardware level** explains how the algorithm is implemented physically, for example, how individual neurons or neural systems perform the computations specified at the algorithmic level.

A2 METHODS OF INVESTIGATION

Key notes

The science of cognitive psychology	Cognitive psychology involves developing theories about human thought processes, deducing hypotheses, and testing predictions based on those hypotheses. Often a hypothesis is tested by several methods to give converging evidence for or against a theory. At least some tests of a theory should be ecologically valid, that is, the test conditions should conform closely enough to real world conditions that we can conclude whether the theory holds true in 'real life'. An additive model of cognition, which assumes serial processing, is often the basis for designing experiments to discover how long a particular cognitive process takes or what happens to task performance when a particular process is prevented.
Research ethics	All research must be ethical. Participants must know what they are volunteering for, must be free to withdraw at any point, must not be caused harm or distress, and must be debriefed at the end of the experiment so they leave with an understanding of what was done and why. Researchers are responsible for ensuring the sound conduct and reporting of research.
Empirical methods	Experimental methods test the effect on cognitive performance of manipulating stimuli or conditions, whereas correlational approaches investigate the relationship between different cognitive functions. Statistical analysis helps distinguish the effects of interest from fluctuations in performance that are due to chance or other variables.
Related topics	Cognition (A1) The history of cognitive psychology (A3)

The science of cognitive psychology

Like the rest of psychology, cognitive psychology is a science. It involves systematically investigating human thought by developing and testing theories. Here are some terms you will encounter in this book:

- A **phenomenon** is the problem or aspect of behavior that interests you. For example, a visual illusion or people's ability to hear someone mention their name above the background noise of a party.
- A **theory** is an integrated set of assumptions and statements about the phenomenon of interest. Theories serve two purposes: i) they **explain** what we already know about the phenomenon, often integrating diverse data sets by postulating general, underlying principles or mechanisms; ii) they enable us to **predict** new information. Theories must be **testable**, that is, they must give rise to specific predictions that can be disproved or **falsified**. Marr's theory of vision is an example (see Topic B4).
- A **paradigm** is the global set of assumptions shared by a community of researchers, including the shared theory or theories, assumptions about

which are the important aspects of the phenomenon under investigation, and assumptions about the best methods for studying the phenomenon. Connectionism (see Topic E4) and cognitive neuropsychology (see Topic A4) are examples. The term 'paradigm' is often used in another sense in cognitive psychology, to mean a research method or technique.

- A **framework** is a general set of principles for understanding a phenomenon or conducting research. It is a useful set of guidelines rather than a testable theory. Craik and Lockhart's levels of processing framework is an example (see Topic D3).
- A **model** is a testable set of propositions about a phenomenon. The terms 'model' and 'theory' are often used interchangeably, though 'model' is usually reserved for explanations with more limited scope. For example, decay theory (see Topic D3) is a claim about general memory function whereas the working memory model (see Topic D2) is a claim specifically about how we temporarily store and process information during complex cognitive tasks.
- A **hypothesis** is a testable proposition, derived from a theory or model.
- A **prediction** is a statement about how the world will behave in a particular circumstance.
- An **experiment** is a means of testing a hypothesis about the effects of certain variables on behavior by manipulating those variables, while holding others constant.
- **Empirical research** or **empiricism** is the building and testing of theories through observation and experimentation (contrast **rationalism**, which is the development of theories from first principles). Cognitive psychology is an empirical science.

As an example of theory development in cognitive psychology, consider the changes in our understanding of short-term memory (discussed in more detail in Topic D2). In the 1960s, Atkinson and Shiffrin proposed a model of memory in which short-term storage of information was separate from long-term storage. They hypothesized that the short-term store functioned as a 'working memory', that is, that it did not merely hold information for brief periods so people could do well in short-term memory tests, but rather that it held information in a form that could be used for complex cognitive activities such as problem solving or reasoning. Baddeley and Hitch tested this hypothesis. They asked participants to remember a string of digits while performing a reasoning task. They predicted that the concurrent digit memory task would impair performance on the reasoning task. In fact, performance on the reasoning task was impaired very little by the digit memory task, much less so than would be expected if both tasks competed for the same limited-capacity store. This finding therefore falsified the hypothesis of a unitary short-term store that functions as a working memory. On the basis of their research, Baddeley and Hitch developed a new model of working memory, comprising separate storage systems (used for remembering the digits in their experiment) and a processing system (used for complex tasks such as reasoning). The new model led to new hypotheses, for example that there are separate visual and verbal storage systems, and new predictions, for example that verbal tasks will interfere more with verbal short-term memory than with visual short-term memory. Empirical tests of these predictions have generally found them to be correct. For example, experiments have shown that repeating 'the the the …' impairs short-term memory for words more than short-term memory for pictures or locations.

The fact that many experiments have supported the predictions made by the working memory model does not mean that the model is true, that it gives an absolutely accurate account of human memory function. An experiment we run tomorrow may still disprove some aspect of the model. Therefore we cannot say that the theory has been proved, merely that it has not been falsified yet. Data that support a prediction are said to **corroborate** the theory, rather than to prove it. Failing to falsify a theory is a useful thing to do – a corroborated theory is better than an untested theory or a falsified theory, and the more experiments we run, the better will be our understanding of the phenomenon and our ability to generate new predictions.

What happens if an experiment produces data that falsify a hypothesis? One option is to modify or abandon the original theory. Another is to question the experiment, to ask whether there is some aspect of the experiment that led to the anomalous result even though the theory may be true. Maybe participants failed to follow the instructions they were given or maybe their preconceived ideas about the topic of investigation distorted their behavior. A solution to this problem of knowing how to deal with anomalous data, and the problem of not being able to prove a theory conclusively, is to test theories in many different ways. **Converging evidence** from a variety of sources helps to strengthen a theory. A theory that has survived a concerted attack from laboratory experiments, neuropsychology, computational modeling and so on is generally considered better than one that has support from only one source. Different empirical methods are considered next, and in Topic A4.

Laboratory-based research enables us to test theories under carefully controlled conditions, so we can be confident that we have taken all the relevant variables and alternative explanations into account. However, for psychology to be an interesting and worthwhile pursuit, it must tell us about how people behave naturally, not just how they behave in the laboratory. It is therefore important that some tests of a theory are **ecologically valid**, in other words that they test whether the theory applies to real-life conditions. To continue with our working memory example, an important aspect of working memory is that it not only stores information for short periods (i.e. that it functions as a short-term memory) but also that it uses that information to support complex cognitive tasks (i.e. it functions as a working memory). Laboratory studies have supported the claim that working memory is involved in a range of complex cognitive tasks, showing mutual interference between tasks that are assumed to require working memory but are otherwise quite different (e.g. problem solving and random number generation; see Topic D2). Research into the use of telephones while driving provides a real-world demonstration of this. Initially, hands-free cellular phone systems were assumed to be safe for use in cars because talking and driving were sufficiently different activities (one verbal, one visuo–spatial) that they should not interfere. However, both tasks also have a working memory component, both requiring monitoring, planning, retrieving information from long-term memory etc. Using a telephone requires keeping track of the conversation, comprehending what is spoken, planning and remembering what to say next, visualizing the other speaker. Driving requires keeping track of one's position relative to other cars, comprehending road signs and monitoring the general situation (e.g. road works, junctions, speed limits), planning maneuvers (when to signal, turn etc.) and recalling one's route. This analysis makes it less surprising that research shows that telephoning impairs driving performance. In a study that predates the current popularity of mobile phones by several decades, Brown, Tickner and Simmonds (1969) found that comprehending and verifying complex sentences impaired drivers' judgments of whether a gap was wide enough to drive through. More recently, Strayer and Johnston (2001) showed that conversations on hands-free cellular phones made people slower to respond to traffic signals in a simulated driving task.

Many controlled experiments, whether laboratory-based or real-world, are based on an **additive model** of cognition. This assumes that if you add a component to a task, then any behavioral changes (in reaction times or error rates) are attributable to the added component. Imagine you are asked to watch a display of lights and your task is to press a key each time a green light flashes. In the first version of the task, every light is green. Let's say your reaction time is 300 msec. That is how long it takes to make a response to a stimulus. In the second version of the task, some lights are red and you press a different key for those. Let's say your reaction time to the green lights is now 500 msec. That is how long it takes to decide which response to make and to make that response. The difference of 200 msec is therefore assumed to reflect the time taken to decide which response to make.

There are several pitfalls with the additive model. One is that it assumes cognitive processes are serial, that we complete one before starting the next. In the example above, we assume that we finish perceiving the stimulus before deciding which key to press. In fact, we may start preparing our response before we are absolutely sure that the light is green (or red). Another problem is that adding or subtracting a process can radically alter a task. For example, researchers of verbal short-term memory (see Topic D2) sometimes ask participants to repeat 'the' aloud while trying to remember lists of words. They assume that their performance on the memory task then reflects how well they can store words in verbal memory without rehearsing them. In practice, however, participants may change their strategy and use visual rather than verbal short-term memory, remembering the written form of the words or forming images of the things they refer to. An increase in errors in the 'repeating the' condition may reflect either the removal of rehearsal from a verbal memory task or a switch to a visual strategy that is less effective for remembering verbal stimuli. This problem is pertinent for cognitive neuropsychology, where brain damage may alter how people tackle tasks rather than simply removing their ability to do one aspect of a task, and cognitive neuroscience, where brain imaging studies subtract brain activity on control tasks from brain activity on the task of interest (see Topic A4).

Research ethics   All research must respect the rights of the research participants. It must avoid causing them harm or distress. Before starting a study, researchers typically submit a description of their planned research to a local research ethics committee, which follows ethical guidelines published by national bodies such as the British Psychological Society and the American Psychological Association. Although cognitive psychology experiments may generally seem fairly innocuous, there are still ethical pitfalls. For example, if you want to see how much people remember incidentally, without trying to, then you may want to give them a surprise memory test. This involves an element of deception, because you cannot warn participants in advance that their memory will be tested. Indeed, you may use a **cover story** (a fictitious account of the aims of the experiment) to distract them from the real purpose of the experiment. At the end of the experiment, it is important to explain to them why the deception was necessary, not least because you do not want them to try to remember every-thing in every subsequent psychology experiment they take part in, just in case there will be a surprise memory test. This explanation at the end of an experiment is called **debriefing**. Some cognitive psychology research can be stressful because it involves looking at distressing images or hearing threatening words. In such cases, participants should be warned that the experiment involves such stimuli. If an experiment involves a manipulation of mood (e.g. see Topic L2), then it is important to reverse that manipulation in case it adversely affects behavior outside the test environment.

Key elements of ethical conduct of research include:

● Ensuring that participants are told enough about the study in advance that they can give **informed consent** to take part in it. In psychological research, this can require careful judgment. Telling people too much about the aims of the experiment can lead them to try to behave in the way they believe the experimenter wants them to behave, rather than behaving 'naturally'. This is known as responding to the **demand characteristics** of the experiment. Telling people too little about the experiment not only prevents them giving truly informed consent, but also encourages them to form their own hypotheses about the aims of the experiment and to behave in apparently bizarre ways to satisfy those aims.

● Informing participants that they may refuse to take part or may withdraw from the study at any time without giving a reason and without adverse effects (e.g. without compromising their treatment in a medical study).

● **Debriefing** participants at the end of the study. Debriefing means explaining why the experiment was conducted, revealing any hidden manipulation or deception, and reversing any manipulations of mood or beliefs.

As in all sciences, psychological research must be scientifically ethical. Scientists have a social duty to report methods and data accurately and truthfully. Most journals require authors to keep their data for 5 to 10 years from the date of publication, so they can be re-analyzed in the event of a query. The authorship of scientific papers includes all (and only) the researchers who have made an intellectual contribution to the conduct or reporting of the research. All authors are deemed responsible for the content of the paper and for the good conduct of the research reported therein.

Empirical methods

One of the difficulties in testing cognitive theories is that we can only do so indirectly. This problem is not peculiar to psychologists. For example, physicists interested in elementary particles cannot observe them directly, but must infer their behavior from their effect on the environment, for instance in a bubble chamber. Similarly, psychologists cannot see cognitive processes in action but must infer their operation from people's behavior. The problem here is that people vary – they differ in their interest in the experiment, in how well they concentrate, in how much they already know about psychology, and a host of other variables. One solution to this problem is to use carefully designed experiments to minimize the influence of individual variation on the measure of interest. Another is to take advantage of the variation by testing whether the behavior you are interested in correlates with another variable that you already know something about, or to test the differences between two very different groups of people.

As an example, imagine you want to test the hypothesis that verbal information is maintained in working memory by a sub-vocal rehearsal process:

- An **experimental** study might involve testing three groups of people. One group simply remembers lists of words for immediate recall. This is a control group; their performance tells us about 'normal' performance on the memory task. A second group tries to remember lists of words and performs a non-verbal task, for example tapping a pattern on a keyboard, at the same time. This is another control group; their performance tells us about memory performance with general distraction. A third group tries to remember the words while performing a task that blocks the hypothesized sub-vocal rehearsal process, such as repeating 'the' aloud. This is the experimental group. Comparing their performance with the two control groups tell us how much repeating 'the' impairs their verbal memory performance, how much of the impairment is due to the general difficulty of having to perform two tasks at once, and how much is due to the specific demands of performing two verbal tasks at once.

  One way of running this experiment is to test every participant in every condition, so each volunteer does the memory test in the experimental condition and both control conditions. This is called a **repeated measures** design because we measure each person's performance repeatedly. The advantage of this design is that it minimizes the effects of individual variation because each person acts as their own control. However, it introduces new sources of variation, for example participants may get tired towards the end of the testing session or they may develop new strategies for remembering the words. An alternative is to use a **between groups** design, assigning a third of participants to each condition. The effects of individual differences are minimized by testing many people in each group and by randomly assigning participants to the different conditions. Because these designs control for individual variation, their results should tell us about cognition in general, not just about the cognitive processes of these individuals. Thus, although many cognitive psychology experiments are conducted on undergraduate psychology students, their results can be extrapolated to people in general.

- A **correlational** study aims to test a sample of people who vary widely. To test the sub-vocal rehearsal hypothesis, one could use a correlational study to discover whether rehearsal speed, that is, speech rate, predicts verbal memory performance. A good sample would be children of different ages because speech rate increases as children get older. Some researchers have done this and found a strong correlation between speech rate and verbal short-term memory, supporting the hypothesis that the faster you can speak, the more words you can rehearse before your memory for them fades. A variation on this study is to compare two groups of people who vary in speech rate, for example younger and older children or people using their native versus a second language.

- With either type of study, we still need **statistical analysis** to tell us what the data mean. In an experimental study, performance in the different conditions will differ simply because people rarely get identical results each time they are tested. We need statistical tests to tell us whether the differences between scores from the different conditions are real, or merely the result of chance fluctuations in performance. In correlational studies, we need statistical tests to tell us the strength of the relationship between speech rate and memory scores.

A3 THE HISTORY OF COGNITIVE PSYCHOLOGY

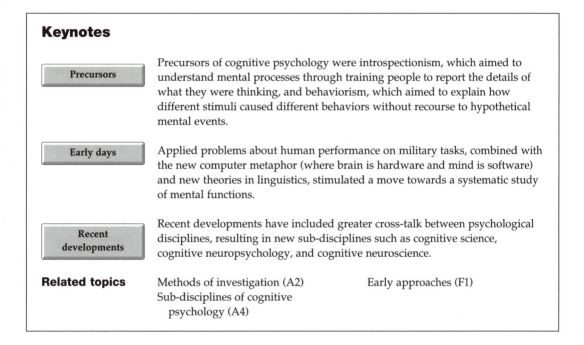

Keynotes

Precursors	Precursors of cognitive psychology were introspectionism, which aimed to understand mental processes through training people to report the details of what they were thinking, and behaviorism, which aimed to explain how different stimuli caused different behaviors without recourse to hypothetical mental events.
Early days	Applied problems about human performance on military tasks, combined with the new computer metaphor (where brain is hardware and mind is software) and new theories in linguistics, stimulated a move towards a systematic study of mental functions.
Recent developments	Recent developments have included greater cross-talk between psychological disciplines, resulting in new sub-disciplines such as cognitive science, cognitive neuropsychology, and cognitive neuroscience.

Related topics	Methods of investigation (A2)	Early approaches (F1)
	Sub-disciplines of cognitive psychology (A4)	

Precursors

In Leipzig in 1879, **Wundt** established the world's first experimental psychology laboratory. His trained observers used **introspection** ('looking into' their minds) to report on their experiences of stimuli. Wundt evaluated their emotional states and measured their reaction times. He emphasized that conscious experiences are more than the sum of their parts, and more than simple passive responses to stimuli. In 1912 Wertheimer extended this approach into **Gestalt psychology**, arguing that mental activity is directed at whole forms rather than component parts of forms, hence *Fig. 1* is typically perceived more quickly as an H and than as an array of Ts (see Topics B4 and B5). Another extension of Wundt's approach was his student Titchener's use of introspection to draw conclusions

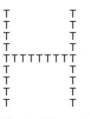

Fig. 1. We perceive the H faster than the Ts.

about the structure of thought, an approach known as **structuralism**. Titchener presented himself as a proponent of Wundt's ideas (e.g. by translating selected parts of Wundt's works from German into English) but his aim of defining the basic components of thoughts conflicted with Wundt's emphasis on whole experiences. The use of introspection fell out of fashion for three reasons: i) new research showed the importance of unconscious, and hence unreportable, processes in vision (Helmholtz) and personality (Freud); ii) unresolvable conflicts between laboratories, such as the **imageless thought controversy** in which trained observers disagreed about whether thoughts were always accompanied by mental images; iii) the increase in interest in the study of mental processes and their functions (rather than structure), known as **functionalism** or **functional psychology** (to distinguish it from the philosophical position). Proponents of functionalism included James, who emphasized the continuous flow of mental processes or 'stream of consciousness', and Dewey, who argued that behavior should be considered as the process of an organism adapting to its environment.

Functional psychology helped push psychology away from studying mental experiences and towards studying behavior. It thus set the scene for **behaviorism**. Behaviorists such as **Watson** and **Skinner** argued that psychology should be concerned only with measuring objectively observable behaviors and with determining stimulus–response relationships, not with studying mental properties such as consciousness. Other behaviorists, such as **Hull** and **Tolman**, included a role for intervening but unobservable mental events in stimulus–response associations.

Early days

Behaviorism dominated North American psychology until the 1970s. Now cognitive psychology is probably the dominant approach in both America and Europe. European interest in studying mental processes for their own sake developed much earlier, stimulated in part by the demands of World War II for a better understanding of **human factors**. The term 'human factors' refers to the human abilities and weaknesses that affect task performance, as opposed to factors such as equipment design. The Unit for Research in Applied Psychology was established in Cambridge UK in 1944 for researching military problems such as pilot fatigue and vigilance in radar operators. When its first director, **Craik**, was asked to develop a model of gun-aiming, he responded by analyzing the problem in terms of the information received by the operator, the processes they had to carry out, and the responses they had to construct. Defense needs, and defense funds, also stimulated later research on selective attention and auditory discrimination, which had implications for applied problems such as those faced by sonar operators trying to distinguish submarines from shoals of fish. A classic example of psychological theorizing from the middle of the 20th century is **Broadbent**'s model of how information is filtered before undergoing further processing and storage (see Topic C1).

Other movements in the 1950s and 1960s helped establish cognitive psychology as a discipline. In **linguistics**, **Chomsky** (see Topic H1) proposed that language is innate, that children are born with a mental grammar that forms the basis for language learning. His proposal of a mental mechanism for understanding language radically and explicitly opposed Skinner's behaviorist theory that we learn language through repeated experience and reinforcement of associations between objects and their names. Developments in computer science helped establish a metaphor for theorizing about cognitive processes as the brain's 'software'. The availability of much greater processing power facilitated developments in statistics, providing new ways of analyzing complex psychological data. **Information theory**, a mathematical approach to explaining the constraints on communication of information, influenced the new theories of cognition that describe the flow of information through the mind.

Recent developments

Although cognitive psychology has been influenced by case studies of the effects of head injury from its early beginnings, and in recent years by computer science, a characteristic feature of the discipline at the start of the 21st century is greater cross-talk with other disciplines. The increasing influences of computational models, systematic studies of brain damage, and new *in vivo* brain imaging techniques have resulted in new sub-disciplines such as cognitive science, cognitive neuropsychology, cognitive neuroscience, social cognitive neuroscience etc. Verbal or qualitative models of information processing are still the foundation stones of mainstream cognitive psychology, but it is increasingly common for them to be supported by evidence from brain-injured patients and quantitative (computational or mathematical) modeling as well as by conventional laboratory experiments. The main sub-disciplines of cognitive psychology are discussed in the next Topic.

A4 SUB-DISCIPLINES OF COGNITIVE PSYCHOLOGY

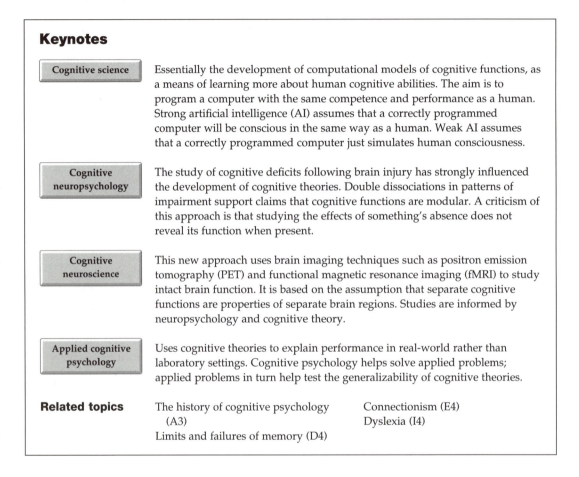

Keynotes

Cognitive science	Essentially the development of computational models of cognitive functions, as a means of learning more about human cognitive abilities. The aim is to program a computer with the same competence and performance as a human. Strong artificial intelligence (AI) assumes that a correctly programmed computer will be conscious in the same way as a human. Weak AI assumes that a correctly programmed computer just simulates human consciousness.
Cognitive neuropsychology	The study of cognitive deficits following brain injury has strongly influenced the development of cognitive theories. Double dissociations in patterns of impairment support claims that cognitive functions are modular. A criticism of this approach is that studying the effects of something's absence does not reveal its function when present.
Cognitive neuroscience	This new approach uses brain imaging techniques such as positron emission tomography (PET) and functional magnetic resonance imaging (fMRI) to study intact brain function. It is based on the assumption that separate cognitive functions are properties of separate brain regions. Studies are informed by neuropsychology and cognitive theory.
Applied cognitive psychology	Uses cognitive theories to explain performance in real-world rather than laboratory settings. Cognitive psychology helps solve applied problems; applied problems in turn help test the generalizability of cognitive theories.

Related topics	The history of cognitive psychology (A3)	Connectionism (E4)
	Limits and failures of memory (D4)	Dyslexia (I4)

Cognitive science

Artificial intelligence (AI) researchers attempt to program computers to replicate human capabilities, for example, robots that can move around their environment and 'see' obstacles in their way. Cognitive science is a marriage between artificial intelligence and cognitive psychology. Cognitive scientists develop computational models of cognitive functions as a means of discovering more about human cognition. They work from the functionalist assumption (see Topic A1) that cognitive functions can be studied without direct reference to the material brain and, indeed, that they can be implemented in any hardware with appropriate computing power. Some take the view known as **weak AI**, that there may be something special about biological brains and sensory systems but that computational models are nonetheless useful tools for improving our theories of cognition. Others hold the **strong AI** position, assuming that a correctly programmed computer will think and be conscious in the same way as a human.

Cognitive scientists aim to program computers with the same **competence** as a human (i.e. the program makes realistic assumptions about how the human brain works) and the same **performance** (i.e. the same pattern of successes and errors on a task). Typically, a qualitative (verbal or flowchart) model of human cognition serves as the basis of the computer model. The computer model is tested by comparing its performance against data from human experimental studies.

Computational modeling is a useful means of testing cognitive theories. Finding that the computer model performs similarly to humans does not prove that the underlying cognitive theory is correct, but shows that the theory gives at least a possible account of the cognitive processes under debate. Finding that the performance of the computer model is unlike that of humans shows that the underlying theory is wrong. Computational modeling is also important because it stimulates better specification of cognitive theories and generates new predictions. For example, conventional accounts of short-term memory describe the need for rehearsal to offset decay of information in temporary storage. However, they do not specify how that rehearsal takes place (how often is each item rehearsed? in what order are items rehearsed?) or how retrieval happens (how do we know where the start of the list is? what happens when we cannot remember an item?). These questions must be solved before a computer can be programmed to mimic accurately human performance on a short-term memory task. A key problem is how, during a serial recall task, we know which item comes next in the sequence. Some solutions to this problem are outlined at the end of Topic D2.

Cognitive neuropsychology

Cognitive theories can be tested by studying the pattern of preserved and impaired cognitive performance after brain injury. If a theory postulates that, say, language production and language comprehension are separate functions, then it should be possible to find people with one sort of brain injury who have difficulty speaking but not comprehending, and people with another sort of brain injury who have difficulty comprehending but not speaking (see Aphasias in Topic H4, and Topics D2 and D4 for other examples). This pattern of data would be an example of a **double dissociation** and supports the assumption of modularity of cognitive functions (see Topic A1). Cognitive psychology and neuropsychology are mutually beneficial – cognitive theories can help explain individual patients' difficulties with particular tasks and, by improving understanding of their condition, help suggest strategies for rehabilitation.

A criticism of cognitive neuropsychology is that studying the effects of something's absence does not reveal its function when present. By analogy, removing sections of track from a railway may cause chaos, but the function of the track is not related to the chaos caused. A difficulty with the approach is the scarcity of people with very localized brain damage that impinges on specific cognitive functions.

Cognitive neuroscience

Cognitive neuroscience is the study of cognitive functions using new imaging techniques such as **positron emission tomography** (PET) and **functional magnetic resonance imaging** (fMRI) to study brain function *in vivo*. These techniques reveal which brain regions are most metabolically active (e.g. which have the greatest blood flow or use the most glucose) when performing different cognitive tasks. The approach assumes that cognitive functions are modular and localized to particular brain regions. Greater metabolic activity in a particular brain region corresponds to greater involvement of that region in the task being performed. Imaging studies of intact brain function are often combined with neuropsychological studies of brain damage to provide converging evidence about the anatomy of particular cognitive functions.

Brain imaging studies generally use subtraction methods, in which brain activity during a control task is subtracted from activity during the task you are interested in. The result of the subtraction shows which areas of the brain are used more by the experimental task than by the control task. For example, if you wanted to know about brain activity while interpreting ambiguous figures such as *Fig. 1* in Topic A1, you would need to subtract brain activity while viewing a non-ambiguous figure, otherwise you would see activity across all the areas involved in visual perception rather than just those involved in the interpretation phase. Choice of an appropriate control task depends on your theory of the processes you are studying, hence brain imaging research is theory-laden. Another problem is that although the spatial resolution of fMRI is good, meaning that it can detect activity in small parts of the brain (PET is less good), its temporal resolution is relatively poor, meaning that it cannot show brain activity related to very rapid cognitive processes. Recent methodological advances have improved temporal resolution by combining fMRI with recording of **event-related potentials** (ERPs), that is, using scalp electrodes to record the brain's electrical responses to discrete stimuli. ERPs have poor spatial resolution but much better temporal resolution than PET or fMRI.

Applied cognitive psychology

From its roots in human factors research (see Topic A3), cognitive psychology has helped solve everyday problems such as how people perform under stress or in noisy environments. The use of cognitive theories to analyze and solve such problems is called applied cognitive psychology. There are mutual benefits, because applying cognitive theories to real-world problems helps to test whether the theories generalize beyond laboratory situations. In other words, it tells us whether people's behavior in the laboratory was similar to how they would behave in 'real life'. Cognitive theories are also influential in applied disciplines such as clinical psychology, where they can tell us about how people's thought processes may have become distorted or maladaptive, and occupational psychology, where they can tell us about how people may cope with learning new work routines or applying information learned on training courses to their everyday work tasks.

B1 SENSATION

Keynotes

Sensation and perception	Sensation is the transduction of energy (e.g. light or sound) into neural impulses. Perception is the interpretation of these impulses to create meaningful internal representations of our environment. Information processing approaches emphasize the influence on perception of top-down processes such as expectations, whereas advocates of direct perception argue that perception is driven solely by data from the environment. Selective attention and forgetting help filter useful information from less useful. Sperling showed that information is lost rapidly from sensory memory.
The visual sensory system	Photoreceptors in the retina convert light energy to neural impulses which are transmitted to the visual cortex via bipolar cells, ganglion cells and the optic nerve. Adaptation to constant stimuli aids detection of change and causes after-effects. Lateral inhibition between cells in a cell layer causes brightness contrast effects and contributes to lightness constancy.
The auditory sensory system	The ear converts pressure changes in the environment (i.e. sound waves) to pressure changes in cochlear fluid. These pressure changes distort the basilar membrane, bending the hair cells attached to it and stimulating auditory neurons.

Related topics

Sensation and perception

Sensation is the first stage of perception. It is a process of **transduction**, the conversion of the energy of light or sound into neural transmission. Perception is the process of interpreting the data thus collected by the sensory organs, to create an internal representation of objects in the environment. Proponents of **direct perception** (e.g. Gibson) argue that we directly perceive the environment from the information it contains, for example perceiving our motion through the world from **optic flow patterns**. In contrast, proponents of the **information processing** approach (e.g. Gregory) argue that perception involves top-down processes of hypothesis formation, in which memory and expectations help us interpret sensory information from the outside world, as well as bottom-up processes of receiving and organizing sensory information. The goal of perception is to identify and categorize objects in the environment, creating meaningful and useable internal **representations** of the external world (see Section E).

We only become aware of a small amount of the information that bombards our senses. If you are concentrating on reading this sentence, you probably will not have noticed the sensation of your clothes against your skin or the buzz of background traffic. Much research and debate has focused on the cognitive processes by which the vast amount of incoming information is reduced to the small amount of important information that we need to function. Even the very early stages of sensation and perception involve filtering incoming light and sound information to extract the most informative portions (e.g. edges in a visual scene). Selective attention processes aid this information filtering (see Section C). Attended information is further reduced by forgetting, leaving meaningful and useful information in working and long-term memory (see Section D).

Sperling's (1960) **partial report technique** provided evidence for the brief persistence in memory of sensory information that has survived perceptual filtering processes. Sperling used a **tachistoscope** (a device allowing brief and accurately timed presentations of stimuli) to present arrays of letters for about 50 msec. Although participants could usually remember only about four of twelve letters in an array, they typically reported seeing more letters. Sperling hypothesized that information about all the letters persisted after the stimulus was removed, but did not last long enough for all the letters to be reported. He tested his hypothesis by playing a tone after presenting the letter array, to tell participants which row of letters to report. When the tone was played immediately after the letter array, participants' performance indicated that around nine letters were available for mental inspection and report, although performance declined dramatically if there was a delay between the letters and the tone. Sperling's findings suggested that detailed visual information persists for about half a second in **iconic memory**. The equivalent auditory store is called **echoic memory** (see Topic D2).

The visual sensory system

Figure 1 shows the structure of the human eye. Light entering the eye through the cornea is focused by the lens to form an inverted image on the retina. Light sensitive cells called **photoreceptors** in the retina respond to information contained in this image. **Rods** can respond to very low levels of light and are most plentiful in the periphery of the retina, hence a faint stimulus visible 'in the corner of one's eye' may become invisible when viewed directly. **Cones** respond to light of particular wavelengths to provide color vision; they are most plentiful in the **fovea**, the region of greatest visual acuity in normal daylight. There are three types of cone, responding to red, green and blue wavelengths of light respectively. Light information is transmitted from the photoreceptors to the **bipolar** and **ganglion cell layers**, thence via the optic nerve to the **lateral geniculate nucleus** (LGN) of the thalamus and the **visual cortex**. Information from the right visual field (the right hand side of the scene in view) arrives at the left hand side of each retina. From there it is transmitted to the left hemisphere of the brain, and rapidly communicated to the right hemisphere via a bundle of nerve fibers called the corpus callosum. Conversely, information from the left visual field arrives at the right hemisphere of the brain slightly before the left.

Sensory systems respond to change, hence new or changing stimuli evoke larger responses than old or unchanging stimuli. The reduction in response to a static stimulus over time is called **adaptation**. Adaptation in the visual sensory system causes **after-images** and after-effects. For example, some cortical cells respond selectively to stimuli of particular width or particular angle of tilt. Adaptation of these cells to parallel tilted lines or 'gratings' causes a **tilt after-effect** when subsequently viewing a vertical grating (see *Fig. 2*). **Lateral inhibition** between cells in a cell layer causes **brightness contrast effects** (see *Fig. 3*)

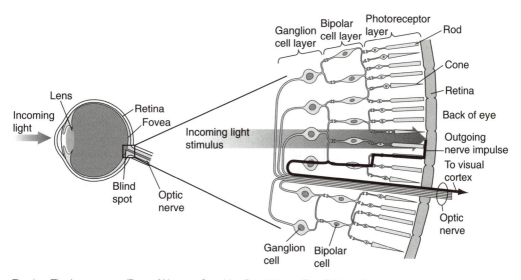

Fig. 1. *The human eye. (Payne/Wenger,* Cognitive Psychology, *First Edition. Copyright © 1988 by Houghton Mifflin Company. Used with permission.)*

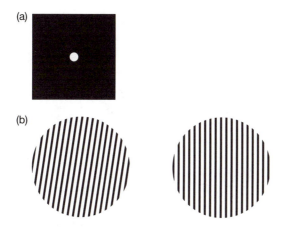

Fig. 2. *Adaptation effects. a) White and black after-images: look at the white dot in the black square for a minute and then at the blank white space next to it – a bright white square should appear with a black dot in it. b) Tilt after-effect: look at the tilted grating for a minute and then view the vertical stimulus; it should appear tilted in the opposite direction.*

Fig. 3. *Brightness contrast effect: the four bars are evenly shaded but appear darker where they meet a paler bar and brighter where they meet a darker bar, hence their edges are accentuated.*

and contributes to **lightness constancy**, that is, our perception of objects as having the same relative brightness even when illumination conditions change.

The auditory sensory system

Vibrating objects cause pressure changes in the fluid (e.g. air or water) surrounding them. These pressure changes are perceived as sounds and can be described as waves having particular amplitude (height) and frequency (number of peaks per second).

Figure 4 shows the structure of the human ear. The **pinna** directs sound waves to the auditory canal, which transmits them to the eardrum or **tympanic membrane**. Vibration of the tympanic membrane causes vibration of three small bones, the hammer or **malleus**, the anvil or **incus**, and the stirrup or **stapes**, which act as levers to amplify the response to the sound. The stirrup transmits the vibration via the oval window to the fluid-filled cochlea. The resulting pressure changes in the cochlear fluid cause the **basilar membrane** to vibrate, bending the **hair cells** that are attached to it. The nerve impulses thus generated are transmitted via the auditory nerve to the auditory cortex of the brain.

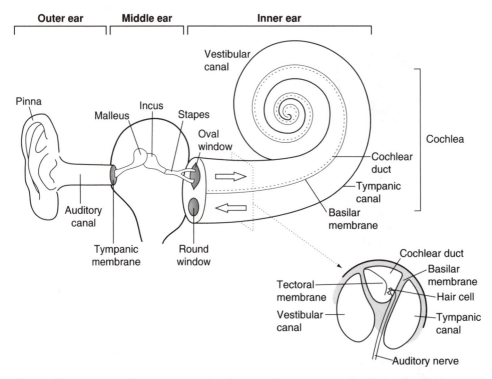

Fig. 4.   The human ear. The arrows show the direction of the sound waves in the cochlear fluid.

B2 AUDITORY PERCEPTION

Keynotes

Pitch perception	The ear provides two types of information about pitch: i) the frequency of firing of auditory neurons responding in synchrony with low frequency sound waves; and ii) the place of maximum distortion of the basilar membrane in response to high frequency sound waves.
Loudness perception	Loud sounds cause more cells to fire than quiet sounds. Some cells only respond to loud sounds.
Sound localization	We can judge where a sound is coming from by comparing its time of arrival, phase differences, and intensity at the two ears.

Related topics	Sensation (B1)	Speech (H4)
	Auditory attention (C1)	Understanding words (I1)

Pitch perception

Pitch is a particularly important aspect of sound information. It helps us discriminate different speakers' voices and distinguish a speaker's mood. High frequency sound waves, that is those that comprise rapid changes in pressure in the fluid that transmits them, are perceived as high pitch. Low frequency waves are perceived as low pitch. High-pitched sounds cause localized vibration of the part of the basilar membrane nearest the oval window of the cochlea (see *Fig. 4*, Topic B1). Low-pitched sounds travel further, causing less well localized vibrations further along the length of the basilar membrane. The **place theory** of pitch perception therefore proposes that sounds with different pitches cause different sets of hair cells to fire, which in turn stimulate different auditory nerves. However, this method of pitch perception probably does not work accurately for very low-pitched sounds, which deform a very broad region of the basilar membrane. Because low-pitched sounds have low frequency waveforms (i.e. relatively slow pressure changes), auditory neurons can respond in synchrony with the peaks of the wave. The **frequency** or **temporal theory** of pitch perception thus proposes that we perceive low-pitched sounds by analyzing the frequency of firing of auditory neurons. These two mechanisms of pitch perception work together, providing complementary sources of information about the auditory stimulus.

Loudness perception

Loud sounds have greater amplitude waveforms than quiet sounds, reflecting greater pressure changes in the air as the sound wave moves through it. Loud sounds cause greater distortion of the basilar membrane, causing more neurons to fire and stimulating some neurons that only respond to very loud sounds.

Sound localization

A sound coming from our left will arrive at our left ear before our right. It will be slightly quieter at our right ear, because of the shadowing effect of our head.

We judge the source of sounds by comparing arrival time and sound intensity at the two ears. Arrival time is more informative when sounds have low pitch; intensity is more informative for sounds with high pitch. For continuous sounds, differences in the phase of the sound waves can also be used. Sound waves are in phase when their peaks happen at the same time, and out of phase when a peak in one wave coincides with a trough in the other. Because sounds from a single source travel further to reach one ear than the other, continuous sounds can produce phase differences in the signals received by each ear.

B3 VISUAL DEPTH PERCEPTION

Keynotes

Monocular cues   Although the visual image on the retina is two-dimensional, it contains cues about depth and distance. Nearer objects appear larger and may occlude more distant objects. Texture gradients and lighting across a scene tell us about the relative positions of objects within the scene. Motion cues (retinal image size changes, optic flow patterns and motion parallax) also give information about distance and are important for accurate interactions between the observer and objects in the world.

Binocular cues   Stereopsis is the process of perceiving an object with both eyes and fusing the slightly different images falling on each retina. It provides powerful depth cues for relatively nearby objects. Depth information may also be provided by the angle of convergence of the two eyes when fixating on an object.

Size constancy   Because the size of the retinal image changes as our proximity to an object changes, accurate judgment of distance is important for size constancy.

Related topics   Sensation (B1)   Visual attention (C2)
                     Visual form perception (B4)

Monocular cues   The image that falls on the retina is two-dimensional, so to perceive the world in three dimensions we must make inferences about the distances between objects and ourselves. We do this using depth cues. **Monocular depth cues** are those available to us when viewing a scene with just one eye. One of these cues is **accommodation**, the extent to which the muscles surrounding the lens contract or expand to alter the shape of the lens. The lens needs to be more spherical to focus on close objects and more elongated to focus on distant objects. The other monocular depth cues are also known as **pictorial cues** because they can be used by artists to give a sense of depth in two-dimensional paintings and drawings.

Interposition is the occlusion of a far object by a nearer one. The predisposition of the visual system to perceive whole forms (see Topic B4) means that we interpret occluded objects as being further away, rather than as being incomplete (see *Fig. 1*).

Objects that are further away produce smaller retinal images than objects that are closer to us. Generally, therefore, the **relative size** of perceived objects gives cues about their distance from us and from each other (*Fig. 2*). However, we also use our knowledge of **familiar size** to infer spatial relationships. Ittelson (1951) showed participants playing cards through a peephole that restricted them to monocular vision and reduced depth cues. Three playing cards were presented individually at the same distance from the observers, who were asked to judge how far away the cards were. Unknown to the observers, the

Fig. 1.   Interposition as a cue to depth: we interpret this picture as a square behind a circle, not a square with a corner cut off balanced on top of a circle.

cards differed in size. The distance estimates show that the observers assumed the cards were the normal size for a playing card and used this familiar size information to judge distance. The card that was twice the normal size was judged to be only half the actual distance away, whereas the card that was half the normal size was judged to be twice as far away as the normal card. **Linear perspective** is a compound cue based on the relative sizes, heights and distances of objects in a scene. Actual or inferred lines linking parallel objects appear to recede towards a point on the horizon (*Fig. 3*). This effect is also seen in *Fig. 2*, where the smaller object is higher up than the larger object, and so seems to be nearer an assumed horizon. In *Fig. 2*, the cues of relative size and linear perspective are working together to make the smaller object seem further away than the larger object.

Elements forming an evenly patterned background to a scene (e.g. pebbles on a beach or blades of grass in a field) appear larger and more distinct when

Fig. 2.   Relative size and linear perspective are depth cues, hence the smaller house appears further away, closer to an assumed horizon.

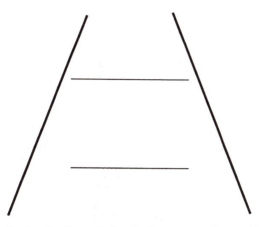

Fig. 3. *The Ponzo illusion: the linear perspective provided by the converging lines provides a strong depth cue, hence the top line appears longer but further away than the bottom line (though in fact they are the same length).*

closer to us, smaller and less distinct when further away. This **texture gradient** provides strong cues to the relative sizes and distances of objects in the scene (see *Fig. 4*).

Lighting and shadow also provide cues to depth. Normally we see objects lit from above by daylight. Ramachandran (1988) showed people an array of ovals shaded from light to dark or dark to light, as shown in *Fig. 5*. Ovals that were light at the top and dark at the bottom were perceived as convex bumps, whereas those that were dark at the top and light at the bottom were perceived as concave dents. This effect can make the craters in pictures of the moon's surface appear as hills rather than depressions. Lighting also provides **aerial perspective**: scattering of light by atmospheric dust makes distant objects

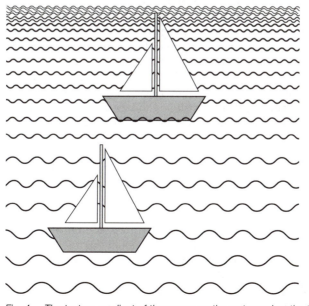

Fig. 4. *The texture gradient of the waves on the water makes the top boat appear more distant and therefore larger than the bottom one.*

Fig. 5. Ramachandran (1988) used stimuli like these to demonstrate the importance of illumination for depth perception. People tend to perceive the stimuli that are lighter at the top as bumps and those that are lighter at the bottom as dents.

appear bluish and indistinct, hence objects with lower contrast appear further away than brightly colored, distinct objects.

Motion cues result from the fact that images of nearby objects move further across the retina than those of distant objects. Objects that are nearer than our fixation point appear to move in the opposite direction to our own movement; objects that are further away move with us. This **motion parallax** can be easily observed from a moving vehicle. A point on the horizon appears to move with us while nearby trees and houses whiz past in the opposite direction. **Optic flow patterns** also provide crucial information about depth. As we move towards an object, the image of the object on the retina gets larger and so do other elements in the scene, for example the elements that provide texture. Optic flow is the name given to this apparent expansion of the visual field as we approach an object. The change in size of the retinal image is however sufficient for us to judge time to contact with an object. Savelsberg, Whiting and Bootsma (1991) demonstrated this by observing the grasping movements made by participants trying to catch balls approaching them. Unknown to the participants, some of the balls were deflating, so they produced less rapidly expanding retinal images than the normal balls. As predicted, participants started to make grasping movements towards the deflating balls later than towards the normal balls, suggesting that they perceived the deflating balls as moving less rapidly towards them.

Binocular cues

Viewing a scene with both eyes gives additional information about depth. One depth cue is **convergence**, the amount that the eyes must point inwards so that each fixates the viewed object at the fovea. Another, more important cue, is **binocular disparity**. For objects less than about 10 m away, each eye receives a slightly different image. These images are fused to give one, three-dimensional representation. This process of perceiving depth through comparison of the two different retinal images is called **stereopsis**.

Size constancy   Although the size of the retinal image changes as we move relative to an object, we perceive objects as having constant size. Depth cues help maintain size constancy; we interpret shrinking retinal images as objects getting further away rather than as shrinking objects. Some visual illusions illustrate the power of depth cues when judging size (see *Figs 3* and *4*). Gregory (1980) argues that the Müller–Lyer illusion is an example of misplaced depth perception (*Fig. 6a*). The fins give the lines the appearance of corners, the fins themselves representing walls or sides coming towards us or receding from us. The line that appears longer is perceived as an inside corner (of a building for example), further away from us than the outside corner represented by the apparently shorter line. Gregory's explanation of such visual illusions emphasizes the role of **top-down**, interpretative processes in perception – the stimuli are illusory because we perceive more than the simple two-dimensional representations provided. However, although the explanation holds for some illusions, it does not give a complete account of the Müller–Lyer illusion, which works even when depth cues are omitted (*Fig. 6b*).

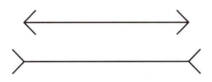

Fig. 6a.   The standard Müller–Lyer illusion. The two horizontal lines are the same length, but the lower one appears longer.

Fig. 6b.   The Müller–Lyer illusion without depth cues. The distance between dots 1 and 2 is the same as the distance between dots 2 and 3, but appears longer (adapted from Zanforlin, M. (1967) Quarterly J. Exp Psychol. with permission of The Experimental Psychology Society).

B4 VISUAL FORM PERCEPTION

Keynotes

Primitive features	Some cells in the visual system function as specialized feature detectors, responding only to particular aspects of stimuli. Performance on visual search tasks confirms that feature analysis occurs at an early stage of visual perception. Stimuli with one type of feature 'pop out' from a background of different features. The binding problem is the problem of explaining how we glue together different features to perceive unitary objects.
Gestalt approach	The Gestalt group of psychologists argued that we perceive forms as distinct from each other and from their background by grouping together elements that are physically close together, follow the same contour, move in the same way or share the same features.
Marr's theory	Marr argued for three major computational stages in form perception, resulting in three different types of representation. The primal sketch represents the main elements of a scene as contours, edges and blobs. The 2.5D representation links these together and uses depth cues to give a viewpoint-dependent representation of the objects in the scene and their relative positions. The 3D representation gives a viewpoint-independent representation of the scene that forms a basis for object identification.
Related topics	Visual depth perception (B3) Gestalt problem solving (F1)
	Object recognition (B5)

Primitive features

Early stages of vision identify features in the visual field. Physiologists Hubel and Weisel won the Nobel prize for identifying cells in the visual cortex that respond to specific features of a visual stimulus. For example, some cells respond to lines tilted at particular angles whereas others respond to lines moving in particular directions. Collectively these cells are called **feature detectors**. In a visual search task participants must detect a target stimulus from an array of distractors (see Topic C2). Anne Treisman used a **visual search task** to show which features were important at a perceptual level, that is, which features formed the building blocks of perception. These so-called **primitive features** include color, orientation, curvature and line intersections. Stimuli having a particular feature stand out easily in arrays of distractor stimuli having a different feature, a phenomenon known as **pop-out** (see *Fig. 1*).

Whereas it is relatively easy to detect stimuli that share no features with their background, it is much harder to detect stimuli that are defined by a specific combination of features that occur in isolation in the background. For example, detecting a blue T against a background of red Ts and blue Xs is a relatively slow and error-prone process. Treisman asked participants to report what they saw when very briefly shown displays containing a red F and green X.

Fig. 1.   Pop-out: The O is easy to spot because its features are different from those of the background Xs. The Y is harder to spot because it shares features with Xs.

Occasionally they reported **illusory conjunctions**, that is they perceived a red X or green F. Treisman's findings suggest that form perception occurs after feature detection (see also Topic C2).

Even though visual perception seems to be based, at a physiological and psychological level, on detection of features, our experience is not of collections of features but rather of cohesive objects. The **binding problem** is the name given to the problem of working out how we combine the products of disparate parts of the visual system (e.g. those that detect edges and those that detect motion) to give a unified conscious experience (e.g. of a moving object). Crick and Koch (1990) argued that binding occurs when neurons responding to the same stimuli fire in synchrony. Storage of percepts in working memory may provide binding at a higher level (see Topic K5).

Gestalt approach

A key problem in form perception is segregating objects from each other and from their background. The Gestalt psychologists argued that we do this by selecting the simplest and most stable arrangements of potentially ambiguous scenes (the **law of Prägnanz**). They proposed several **principles of perceptual organization**:

- The **principle of proximity**: elements that are close together tend to be perceived as belonging to the same object. Hence we perceive *Fig. 2a* as four rows rather than four columns.
- The **principle of similarity**: similar elements tend to be grouped together, hence we perceive *Fig. 2b* as a black 'V' against a white background rather than as an even array of dots.
- The **principle of good continuation**: we tend to perceive smooth curves rather than sudden changes of direction, hence we see *Fig. 2c* as two gently curving lines that cross over each other, rather than as a seagull looking at its reflection.
- The **principle of closure**: we tend to fill in missing parts of figures to see complete forms, hence *Fig. 2d* is seen as a square despite the blank corners.
- The **principle of common fate**: elements that move together are perceived as belonging to the same object, hence we could spot a camouflaged animal if it moved or pick out a stationary person in a moving crowd.

The principles of good continuation and closure combine to give us a preference for perceiving whole objects even when none exists. A possible consequence of this preference is the phenomenon of **subjective contours** (see *Fig. 3*).

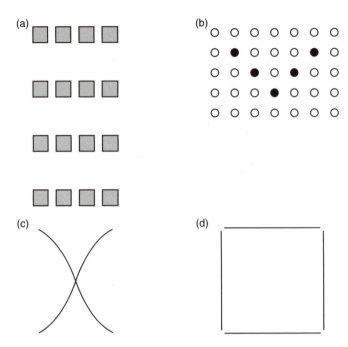

Fig. 2. Gestalt principles of perceptual organization: (a) proximity, (b) similarity, (c) good continuation, (d) closure.

Fig. 3. Subjective contours: Kanizsa's triangle is perceived as a complete, bright white triangle in the foreground of the figure. Reprinted from Kanizsa (1976) Scientific American, with permission

The Gestalt psychologists argued that the organization of perception mirrored the organization of electrical fields in the brain. However, there was no evidence to support this doctrine of **isomorphism**. The Gestalt principles remain useful descriptions of form perception but not an explanation of it (see also Topic F1).

Marr's theory

Marr's (1982) theory of vision is important because it specifies the problems that must be solved by the brain to distinguish visual features in the environment, and it offers solutions to those problems. Specifically, Marr identifies three stages of computation performed on the light stimulus hitting the retina. The resulting types of representation are as follows:

- The **primal sketch** is the product of early visual processes that analyze changes in light intensity in the visual scene. Shifting patterns of illumination mean that absolute levels of light intensity are often not very informative, therefore intensity changes are analyzed across groups of points (or pixels) in the scene. Areas of maximal intensity change are known as **zero crossing segments**. They are represented as edges, blobs, bars and terminations which together make up the **raw primal sketch**. In the **full primal sketch**, the elements of the raw primal sketch are grouped together using principles similar to those proposed by the Gestalt theorists, hence the full primal sketch contains two-dimensional information about contour and form.
- The **2.5D sketch** is a viewer-centered representation of the orientation and depth of visible surfaces. It is constructed from the primal sketch using depth cues such as texture, perspective and relative motion. Because it is viewer-centered, a 2.5D representation of the same scene will vary according to where it is viewed from. It therefore provides a poor basis for object recognition. The name reflects the fact that it contains more information than the two-dimensional primal sketch but is not three-dimensional.
- The **3D sketch** is an object-centered representation. It represents objects in three-dimensional space independently of the observer's viewpoint and thus includes information about surfaces hidden from view.

B5 OBJECT RECOGNITION

Keynotes

The problem of object recognition	To recognize an object, we must create a three-dimensional representation from the two two-dimensional images falling on the retinas, then identify and name that three-dimensional shape by referring to information stored in semantic memory. Studies of people with visual agnosia suggest that object recognition and classification occur at a later stage of visual processing than feature detection.
Pattern recognition	Template theories propose that two-dimensional patterns, such as the letter A, are recognized by matching the visual stimulus to a template stored in memory. Feature theories, for example Selfridge's pandemonium model, propose that we compare sets of features (rather than complete forms) against the features of prototypes in memory. Template theories are too simplistic to account for our ability to recognize varied and novel patterns. Feature theories ignore evidence that processing of global form often takes priority over processing of local features and is sensitive to the context in which the stimulus appears.
Object recognition	In Marr's theory, the 3D sketch provides the basis for object identification. It is created from cylinders constructed around the main axes of the 2.5D representation, which are relatively insensitive to viewpoint and thus a good basis for creating a viewer-independent representation. In Biederman's recognition-by-components theory, a three-dimensional representation is constructed from geometric units called geons. Thirty-six geons are sufficient to describe a vast number of objects.
Scene perception	Studies of change blindness and inattentional blindness show that people can fail to notice large parts of visual scenes if they are outside the region of focal attention. Despite this, we usually feel that we have a veridical representation of the world, that we perceive everything we are looking at. This discrepancy can be explained by assuming that we use top-down processes to construct representations from limited sensory input.
Related topics	Visual form perception (B4) Face recognition (B6)

The problem of object recognition

The image falling on each retina is two-dimensional and each encounter with a particular object gives a different image that depends on our viewpoint. Object recognition is an interesting perceptual problem because we must use these unique, two-dimensional images to work out the three-dimensional shape of the object. For example, an open door produces a trapezoid-shaped image yet we recognize the image as that of a rectangular door. Having established the three-dimensional shape of the viewed object, we must then classify it on the basis of our knowledge of objects. We are able to do this even though we have encountered many different instances of doors, some of which may not even be rectangular.

Humphreys and Bruce (1989) proposed a three-stage model of object recognition:

- **Perceptual classification** is the matching of visual information about the viewed object with a **structural description** of that object in memory, that is, a description of the components of the object and their arrangement.
- **Semantic classification** is recognition of the object as something that relates to the world in a particular way and serves particular functions. For example, our semantic memory (see Section D) may include the information that doors are often found in houses and serve to seal entrances and exits.
- **Naming** is retrieval of the object's name from memory.

Neuropsychological case studies provide support for claims that object recognition occurs after early visual processes such as feature detection. People with **visual agnosia** have normal vision, in that they can perceive colors and movement, but cannot identify objects or even simple shapes such as letters or circles, which might suggest that they cannot organize incoming visual information to perceive complete forms. Although some people with agnosia can copy drawings, they appear to do this on a feature-by-feature basis. It is debatable whether agnosia is better considered a problem with form perception or a problem with semantic classification. Most people with agnosia can give accurate definitions of objects they cannot recognize (e.g. they can say what a 'carrot' is yet not recognize a carrot when shown one), suggesting that some semantic knowledge is intact. The problem appears to be with linking perception to semantic memory. However, the fact that agnosia is sometimes limited to particular domains, such as words, faces or living things, suggests a problem with a subset of semantic memory rather than a global problem with form perception (see Topic D4).

Pattern recognition

Theories of pattern recognition focus on the problem of identifying simple two-dimensional shapes such as letters and numbers from sets of features. **Template theories** suppose that perceived forms are normalized, that is the representation is adjusted to a conventional position and size, and then matched with a template in memory. Template theories have been used to develop text-recognition machines that work well in limited domains (e.g. with typed letters), but they provide a rather poor account of our impressive ability to recognize a wide variety of scribbles and fonts as particular letters. It is inconceivable that we might have different templates for each letter in each font we might encounter, yet unclear how normalization processes could cope with such a varied input.

Feature theories suppose that we recognize patterns by identifying their component features and comparing these features against the features of **prototypes** in memory. Neisser (1964) provided some empirical support for feature theories, by showing that it takes longer to detect letters when they are surrounded by distractors with similar features than when the distractors are dissimilar (e.g. it takes longer to spot the O in GQDOCQ than in ZMXOTL).

Selfridge's (1959) **pandemonium model** is a type of feature theory. According to this model, **word recognition** is accomplished in four stages by 'demons': (i) image demons record the retinal image; (ii) feature demons each look for a particular feature in this image record and respond if they find it; (iii) cognitive demons look for evidence of a particular set of features in the responses of the feature demons, responding most strongly if the entire set of features is present; (iv) a decision demon observes the ensuing pandemonium, looking for the cognitive demon that responds most strongly and deciding that the stimulus in the real world must correspond to the target that the most active cognitive demon was looking for. Translated into the terminology of modern computational modeling, 'demons' can be thought of as nodes in a neural network and their 'responses' as the activation level of the nodes exceeding threshold and thereby passing activation up to the next layer of nodes in the network (see Topic E4). Word recognition is addressed in more detail in Topic I1.

Feature theories are consistent with the physiological evidence for feature-detecting cells in the visual system, and the evidence from visual search tasks for primitive features (see Topic B4), but they emphasize bottom-up processes at the expense of **top-down processes** that are also known to be important. For example, Navon (1977) showed participants stimuli like that illustrated in *Fig. 1* and asked them to identify either the large letter or its component small letters. Identification of the large letter, that is, the **global feature**, was relatively fast and unaffected by the identity of the small letters, whereas identification of the small letters or **local features** was slower, particularly if the identity of the small letters differed from that of the large letter, as in *Fig. 1*. Findings such as these suggest that form perception takes priority over feature perception. **Context effects** also demonstrate the influence of top-down processes. The letter string in *Fig. 2* is likely to be read as 'grass-roots' even though the 'g' and 'o' characters are identical. Neither feature nor template theories are well-equipped to explain recognition of complex, three-dimensional objects or our ability to recognize the likely function of novel objects.

```
GGGGGG
G
G
G
GGGG
G
G
G
G
```

Fig. 1.   *Global and local features: People tend to be faster to identify the global feature (F) than the local features (G).*

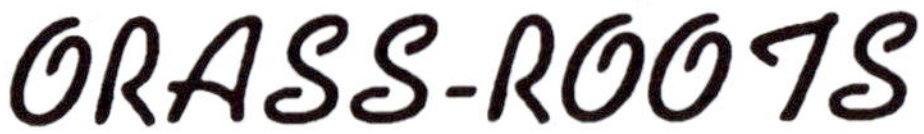

ORASS-ROOTS

Fig. 2.   *Context effects in object recognition: the 'G' and the 'O's are the same character.*

Object recognition

In **Marr's theory** of vision (see Topic B4), the 2.5D sketch provides a relatively sophisticated representation of the visual input. However, it only includes the visible surfaces of objects and therefore is an inadequate basis for identifying those objects. The **3D sketch** includes information about surfaces hidden from view and thus provides the basis for object recognition. Marr and Nishihara (1978) proposed that the 3D sketch is constructed by identifying the major axes of the object. Identification of the main axis of an object is less sensitive to viewpoint than identification of shape or size. It is aided by identification of concavities, points where the object's contours bend inwards. The 3D sketch is built up from cylinders constructed around the main axes of the object.

In his **recognition-by-components theory**, Biederman (1987) proposed that we recognize objects as combinations of simple three-dimensional components called **geons**. A set of only 36 geons is sufficient to generate an enormous number of object prototypes (see *Fig. 3* for an example). Like Marr and Nishihara, Biederman argued that identifying concavities is an important step in working out which geons a particular object comprises. Biederman also argued that we identify **non-accidental properties** in the stimulus, that is, properties that vary little with viewpoint. Non-accidental properties include parallel lines, symmetry and curvature. Having determined the component geons in a stimulus, we match that arrangement of geons with geon-based object representations stored in memory.

Fig. 3.   An object prototype made from geons.

Scene perception

Studies of **change blindness** (failure to notice a change in a scene) and **inattentional blindness** (failure to perceive part of a scene) show that our apparently complete perceptions of complex scenes are actually somewhat illusory. In a typical change blindness task, participants are asked to focus on a particular aspect of a scene. Another aspect of the scene changes and participants are asked if they detected the change. Typically, many fail to do so. Although it is perhaps unsurprising that we fail to notice changes in peripheral details, such as people's clothes or background scenery, there have been some demonstrations of remarkable failures to notice gross changes. For example, Simons and Levin (1998) showed that we are vulnerable to change blindness in real life as well as in the laboratory. The experimenter stopped pedestrians and asked them for directions to a place on a map he was holding. Their conversation was interrupted when the experimenter stood aside to let two men carrying a door pass by. While the door blocked them from the pedestrian's view, the experimenter changed places with one of the accomplices carrying the door. The accomplice then continued the conversation with the pedestrian. Half the pedestrians failed to notice that they were now talking to a different person, even though the experimenter and his accomplice looked quite different.

Inattentional blindness also occurs for surprisingly large components of visual scenes. In a replication and extension of an experiment on attention by Neisser and Becklen (see Topic C2), Simons and Chabris (1999) showed people a video of a basketball game and asked them to count how many passes one of the teams made. Part way through the game, someone dressed in a gorilla suit walked right across the scene. Across a range of conditions, an average of only 43% of participants noticed.

Change blindness and inattentional blindness studies show that we perceive little outside the immediate focus of our attention. However, we generally feel that our perception is far more accurate than this, for example that we notice the appearance of the person we are currently talking to. This discrepancy between the experimental data and our experiences illustrates the role of top-down 'filling in' processes in perception. In other words, much of what we perceive comes not from sensory input from the world but rather from our use of expectations and knowledge to construct scenes on the basis of a rather limited amount of attended sensory input.

B6 FACE RECOGNITION

Keynotes

Face recognition is special	Although face recognition shares many features with general object recognition, it also differs in important ways. The precise configuration of features is important. We are exceptionally good at recognizing upright faces but not upside-down faces. Selective attention to faces seems to be innate. Face recognition can be selectively damaged, leaving other aspects of object recognition intact.
Models of face recognition	Models focus on the interaction between face recognition and other aspects of person identification such as naming and knowing where we have met them before. Bruce and Young (1986) proposed sequential processing stages passing from face recognition units (FRUs) to person identity nodes (PINs) to name generation. This model cannot explain covert recognition in prosopagnosia, but an updated connectionist version (Burton, Bruce and Hancock, 1999) can do so. This model comprises pools of nodes corresponding to FRUs, PINs, semantic information units and name recognition units. When shown a familiar face, activation of the appropriate FRU and PIN in turn activates semantic information units, which then activate PINs for people sharing similar characteristics (domicile, occupation etc.).
Related topic	Object recognition (B5)

Face recognition is special

There are several reasons for considering face recognition to be a special type of object recognition:

- Faces all share the same basic features (eyes, nose, mouth etc.) arranged in the same way, yet we are exceptionally well able to discriminate different faces and to identify different emotions expressed in a single face.
- Face recognition is particularly sensitive to orientation. It is much easier to recognize an upright face than an inverted face, and the difference in ease is greater than for other types of object.
- Face recognition seems less dependent on edge detection, and more dependent on light and shade patterns, than other object recognition. Thus it is very difficult to recognize familiar faces from photographic negatives.
- Faces are more than the sum of their parts; the precise configuration of features is important for face recognition. The so-called Thatcher illusion (Thompson, 1980; see *Fig. 1*) illustrates the importance of **configurational information** for perception of upright but not inverted faces. Young, Hellawell and Hay (1987) provided another demonstration of this. They cut pictures of famous people in half horizontally and asked participants to name the top half of each face. When the faces were upright, naming was much harder when the top half was aligned with the wrong bottom half

than when it was viewed alone. Naming of the mismatched faces actually improved when they were shown upside down.

- Newborn infants have a preference for looking at faces rather than other stimuli, suggesting that attention to faces is an innate predisposition.
- People can suffer from selective deficits in face recognition, called **prosopagnosia**, while still being able to recognize objects and identify people by voice or gait.

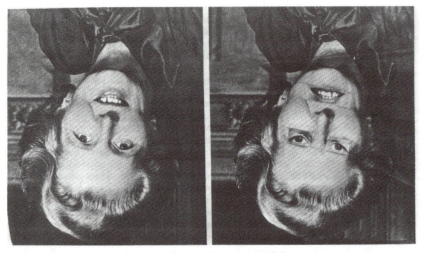

Fig. 1.   The Margaret Thatcher illusion. Turn the page upside down to see the difference between the two pictures. (Reproduced with permission from Thompson, P. Margaret Thatcher: a new illusion (1980), 9, 483–484, Pion, London.)

Models of face recognition

Bruce and Young (1986) proposed an influential model of face recognition that was based on several earlier models. The model assumes that early visual processing of faces proceeds as it does for objects. A representation of a familiar face then triggers activation in a **face recognition unit** (FRU), which contains information about the appearance of a particular person. Activation of a face recognition unit in turn activates appropriate **person identity nodes** (PINs), which contain information about the person (e.g. their occupation, where they live). The final stage is **name generation**. Evidence for name generation being a separate, late stage comes from patients who are able to recognize that a face is famous and give biographical information about that person, yet cannot retrieve their name. Many of us experience a similar problem in everyday life, when we know that we know someone yet have their name 'on the tip of our tongue'. Bruce and Young's model also includes modules for analyzing speech and facial expression.

Bauer (1984) reported **covert recognition** in patients with prosopagnosia. Although the patients appeared to be unable to recognize faces, they showed a galvanic skin response to the familiar faces (a galvanic skin response is a change in the conductance of the skin, usually as a result of autonomic responses to threat but also seen in response to familiar stimuli). This finding is problematic for Bruce and Young's model because it suggests some residual but unconscious face recognition, a possibility not allowed for in the model. Young, Hellawell and de Haan (1988) examined covert recognition further. They asked a prosopagnosic patient known as P.H. to decide whether people's names were familiar to him. P.H. was faster at deciding that a name was familiar when it was preceded by the face of a related person than when it was preceded by an unrelated face. This facilitation, **semantic priming,** occurred even though he was unable to recognize the faces themselves. It suggests that he retained some quite specific information about person identity, even though he could not use it overtly to say who the faces belonged to. This finding is problematic for the model because person identity nodes should only be activated by the output from the appropriate face recognition unit, yet P.H. seemed unable to recognize faces.

Burton, Bruce and Hancock's (1999) computational model tackles both the perceptual and cognitive aspects of face recognition. The perceptual component uses **principal components analysis**, an image-processing technique that preserves information about light and shade patterns rather than just edges. The cognitive component is an **interactive activation and competition model** (IAC model). It is a connectionist architecture comprising sets of processing units. Bi-directional links between units in different sets are excitatory, whereas links within a set are inhibitory. FRUs and name recognition units connect to PINs containing information about who the person is, hence someone can be judged familiar on the basis of seeing their face or reading their name. The PINs connect to **semantic information units** (SIUs), which contain information about occupation, hobbies etc. (see *Fig. 2*).

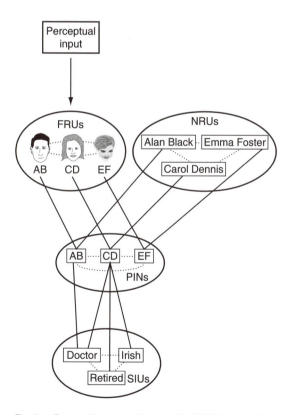

Fig. 2. Burton, Young and Hancock's (1999) model of face recognition. The solid lines are excitatory links, the dotted lines are inhibitory links. Alan Black (AB) and Carol Dennis (CD) are both doctors. When the observer sees CD's face, activation of CD's face recognition unit (FRU) inhibits activation of other FRUs and activates the person identity node (PIN) for CD. CD's PIN in turn activates semantic information about CD, including the fact that she is a doctor. Because AB is also a doctor, some activation from the doctor node passes back to the PIN for AB, hence seeing CB's face would speed recognition of AB.

The IAC model can explain semantic priming effects, in which seeing someone's face or name facilitates recognition of a related person. Recognition of the first face or name activates the SIUs for that person. Some of these will be shared by the related person, so activation feeds back to the PIN for that person. The model accounts for covert recognition in prosopagnosia by assuming FRUs that can be activated normally but are only weakly connected to the rest of the person recognition system. When a face is encountered, its FRU only weakly activates the corresponding PIN. This weak PIN activation then activates SIUs which, because their links are bi-directional, in turn activate related PINs and name recognition units, causing priming effects such as those shown by P.H.

C1 AUDITORY ATTENTION

Keynotes

Selective auditory attention	Mechanisms of selective attention allow us to concentrate on one auditory message while ignoring other messages that differ in terms of speaker's voice and location, thus solving the so-called cocktail party problem. Our ability to hear someone mention our name even when we are attending to another conversation suggests that even the ignored message or conversation undergoes some processing. Dichotic listening experiments have explored the extent of this processing.
Theories of selective attention	Early research suggested very little processing of the unattended message in dichotic listening experiments, leading Broadbent to argue that selection occurs early, at the level of sensation rather than perception. Tests of Broadbent's model revealed more extensive processing of the unattended message, leading to proposals of late selection or a flexible bottleneck in processing. Jones has emphasized the effect of the nature of the distracting material on our ability to selectively attend, arguing that attention involves maintaining target and distractor information in separate streams.

Related topics   Sensation (B1)   Divided attention and dual task
                           Auditory perception (B2)   performance (C3)

Selective auditory attention

Cherry (1953) used simultaneous presentation of two messages to investigate the problem known as the **cocktail party phenomenon**. This is the problem of how we are able to follow one conversation while ignoring other conversations going on around us. We are not able to close our ears to the unwanted conversations, so we must be able to select the important information in some way and filter out the rest. This ability to focus on some aspects of the environment (auditory or visual) and ignore others is called selective attention.

Cherry found that we use the physical characteristics of the stimuli, such as the intensity and source of a message, to separate the wanted message from the unwanted. When a different message was presented to each ear in the same voice, listeners found it almost impossible to separate the two messages.

Another aspect of the cocktail party phenomenon is that salient information breaks through the attentional filter. Thus we may hear someone mention our name even though we have ignored the rest of their conversation (Moray, 1959). Theories of attention must therefore explain **attention switching** as well as selective attention.

In **dichotic listening** experiments, participants are asked to shadow (i.e. repeat aloud) the message reaching one ear, while ignoring the message reaching the other ear. Several influential theories of selective attention are based on findings from dichotic listening experiments exploring the extent to which the unattended message is processed. The different theories make different claims about when the unwanted information is filtered out (see *Fig. 1*).

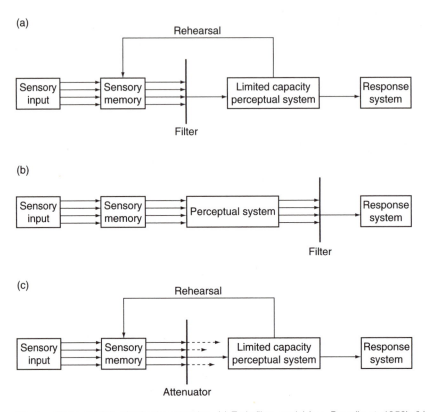

Fig. 1.   Three models of selective attention. (a) Early filter model (e.g. Broadbent, 1958); (b) late filter model (e.g. Deutsch and Deutsch, 1963); (c) attenuation model (e.g. Treisman, 1964).

Theories of selective attention

Early dichotic listening experiments suggested very little processing of the unattended message, thus listeners did not notice if the unattended message was spoken in a foreign language and could not recall words repeated many times in that message. If listeners were asked to attend to both messages, they usually reported the incoming stimuli ear by ear, rather than reporting them in order of input. For example, if one ear received the stimulus '739' and the other heard '256' at the same time, people were more likely to report the entire message as '739256' rather than as '723596'. On the basis of such findings, Broadbent (1958) proposed an **early filter model**. He argued that incoming sensory stimuli gain access in parallel to a sensory buffer (the S-system in his model, akin to the concept of echoic or iconic memory, see Topics B1 and D2). Only one of the competing stimuli gets through the selective filter, thus avoiding overload of limited capacity perceptual processes (the P-system).

Broadbent's model stimulated tests that eventually falsified it. For example, Underwood (1974) showed that with practice at dichotic listening people got much better at reporting details of the unattended message. Practice reduces the amount of processing resources that must be devoted to the shadowing task, so Underwood's finding suggests that filtering can occur at different stages depending on how much processing capacity is available. Allport, Antonis and Reynolds (1972) showed good memory for both messages if they are presented in different modalities, as spoken words and pictures rather than as two spoken messages, suggesting later filtering (see Topic C3). The observation that we tend to hear our name mentioned even in unattended speech suggests some processing of the unattended stimulus. Von Wright, Anderson and Stenman (1975) provided further evidence of continued semantic (meaning-based) processing by demonstrating galvanic skin responses (see Topic B6) to words that were originally presented in the unattended message.

On the basis of findings like these, subsequent theorists suggested that filtering occurs later in the cognitive system. In their **late filter model**, Deutsch and Deutsch (1963) argued that all stimuli are processed until the point at which we need to select which stimulus to respond to. Treisman's (1964) **attenuation model** incorporates three processing levels, physical processing (e.g. processing sound properties such as pitch and intensity), linguistic processing (e.g. segmenting the sound stream into words), and semantic processing (analyzing meaning). There is a flexible bottleneck in processing, depending on available processing capacity. Information is only filtered when there are insufficient resources to process it further and other information has priority. The filtered information is attenuated, that is it receives less processing, rather than being completely unprocessed. Theories that assume late selection can deal better with our ability to switch attention, because even unattended stimuli receive sufficient processing to be identified as salient (e.g. our name) and capture further processing resources if appropriate given the task demands. **Johnston and Heinz** (1978) also argued for flexibility in selection. They suggested that early selection requires fewer mental resources (see Topic C3) than later selection, hence the point at which selection occurs will depend on the amount of resources available.

Jones (e.g. Jones *et al.*, 1999) has explored the influence of the type of distraction on our ability to selectively attend to a particular aspect of the environment. Specifically, he has investigated the effects of different types of background noise on performance of short-term memory (STM) tasks. Jones' **changing state hypothesis** proposes that the degree of change in the background noise determines the extent to which it disrupts performance on the STM task. Evidence in support of this hypothesis includes findings that a constant background tone does not interfere with STM unless it is uncomfortably loud, whereas a changing or pulsed sound does interfere. Rhyming words (tea, sea, me) interfere less than non-rhyming words (bat, sun, toe). Jones argues that selective attention involves **streaming** the target and distractor information, that is, perceiving them and maintaining them in memory as two separate, temporally-ordered sequences of events. Stimuli that form a more coherent stream, such as a constant tone or rhyming words, are easier to keep separate from the stream of target words than those that change frequently. Streaming happens unconsciously at the level of sensation and perception, as well as consciously at the level of deliberately maintaining items in their correct serial order for a STM task (see Topic D2).

C2 VISUAL ATTENTION

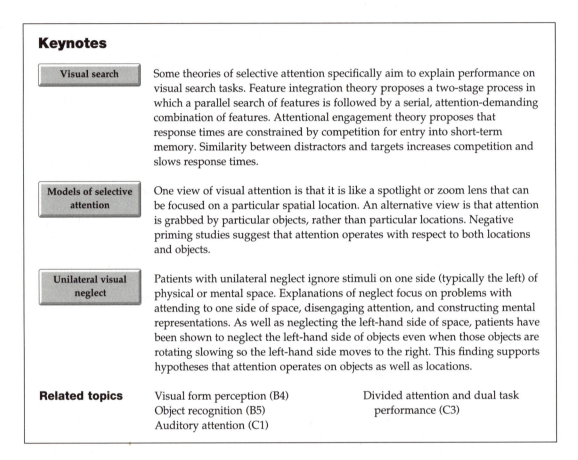

Keynotes

Visual search	Some theories of selective attention specifically aim to explain performance on visual search tasks. Feature integration theory proposes a two-stage process in which a parallel search of features is followed by a serial, attention-demanding combination of features. Attentional engagement theory proposes that response times are constrained by competition for entry into short-term memory. Similarity between distractors and targets increases competition and slows response times.
Models of selective attention	One view of visual attention is that it is like a spotlight or zoom lens that can be focused on a particular spatial location. An alternative view is that attention is grabbed by particular objects, rather than particular locations. Negative priming studies suggest that attention operates with respect to both locations and objects.
Unilateral visual neglect	Patients with unilateral neglect ignore stimuli on one side (typically the left) of physical or mental space. Explanations of neglect focus on problems with attending to one side of space, disengaging attention, and constructing mental representations. As well as neglecting the left-hand side of space, patients have been shown to neglect the left-hand side of objects even when those objects are rotating slowing so the left-hand side moves to the right. This finding supports hypotheses that attention operates on objects as well as locations.
Related topics	Visual form perception (B4) Divided attention and dual task Object recognition (B5) performance (C3) Auditory attention (C1)

Visual search   Visual search tasks require detection of a particular stimulus, called the 'target', from a visual array of other stimuli, called the 'distractors'. Researchers can test theories about the basic features and processes used by the visual system by varying the similarity between the targets and distractors, and measuring the speed with which people can detect the target.

    Treisman's (1988) **feature integration theory** aims to explain performance on visual search tasks. For example, people are quick at detecting a stimulus such as an O against a background of stimuli with different basic features, such as Xs. However, they are slow at detecting stimuli that share features with the background distractors, for example a red O against a background of blue Os and red Xs. Observers occasionally report **illusory conjunctions** of features, for example they may perceive a red O when only blue Os and red Xs were presented (see Topic B4). Feature integration theory assumes a two-stage search process:

- Feature processing. Feature processing is a parallel process, that is, all the features in the display are processed simultaneously. This stage is therefore fast.

Recognizing an O against a background of Xs can be completed at this stage and, because the search is parallel, the O is detected as quickly if it is embedded in a large array of Xs as when it is sandwiched between just two Xs.

- Feature combination. Individual features are combined into objects. This stage is slower because it is serial, that is, the stimuli are processed one at a time. Feature combination requires attention to bind the features together. Stored knowledge of possible objects also contributes. People experience illusory conjunctions when there is insufficient attention (e.g. when attention must be spread over a wide range of locations rather than focused on a particular location) and stored knowledge cannot support performance (e.g. when arbitrary feature combinations are used, e.g. differently colored letters).

Duncan and Humphreys (1992) offer an alternative explanation of similar experimental findings. Their **attentional engagement theory** proposes that visual search times depend on the degree of similarity between target and distractor stimuli, and the similarity between the distractors. Following a parallel, feature-processing stage, there is competition for entry of stimuli into visual short-term memory (STM). Selective attention is a product of this competition phase. Unique targets enter STM unimpeded, but those that are similar to distractors have to compete to enter STM. This competition phase is easier if the distractors are all the same than if they are different. People are slow at detecting a red O against a background of blue Os and red Xs because the distractors share features with the target stimulus and there are different sorts of distractors. In *Fig. 1*, it is relatively easy to spot the upside-down T because the distractors are all identical. Feature integration theory should predict greater difficulty with this task because the targets and distractors share identical features.

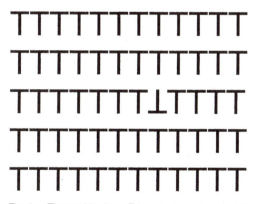

Fig. 1.   The upside down T is easy to spot against the uniform background, even though its features are identical to those of the background elements.

Models of selective attention

A common view of visual attention is that it operates like a **spotlight** (e.g. Posner, 1980) or zoom lens that ranges over representations in mental space. The spotlight can be focused on a particular spatial location and enhances processing of stimuli falling within its beam. Posner's **dot probe studies** support the idea that attention is directed to particular locations rather than to particular objects. Posner asked participants to fixate their gaze on a point in the center of a screen and to respond as quickly as possible whenever a dot appeared on either side of the fixation point. On some trials, the appearance of the dot was preceded by an arrow that either correctly or misleadingly predicted the location of the dot. When the arrow correctly pointed to the location in which the dot would appear, response times were faster than when there was no cue to the dot's location. When the arrow pointed to the side opposite that in which the dot would appear, response times were slowed. Response times overall were too fast to suggest that participants were moving their gaze away from the fixation point. Rather, Posner argued that they made **covert attention shifts**, that is, they shifted their 'mental spotlight' without moving their eyes.

The results from a study by Neisser and Becklen (1975) are difficult to explain with a spotlight model of attention. They asked participants to report information about one moving scene while ignoring another moving scene superimposed on it. Participants were surprisingly well able to do this, suggesting that they could selectively attend to one set of stimuli in a particular location and ignore another set, even though both sets of stimuli should be equally illuminated by an attentional spotlight focused on that location. Some authors, for example Duncan (1984), have therefore argued that attention is directed at particular objects rather than at particular locations. Juola and colleagues (1991) provide further evidence for **object-based models** of attention. They asked participants to detect stimuli appearing in concentric rings. If they were cued to expect a stimulus to appear in an outer ring, then performance was better for stimuli in that outer ring than for stimuli in inner rings. A spotlight model of attention would predict that performance would always be good for the inner rings because they fall in the center of the spotlight, yet Juola's results show that participants could attend selectively to an object (the outer ring) while ignoring stimuli surrounded by that object.

Negative priming studies (e.g. Tipper, 1985) suggest that attention may be directed at both objects and locations. They suggest that selective attention involves active **inhibition** of unwanted information, as well as selection of wanted information. They also show that unattended objects are processed to quite a high, semantic level even though they are ignored (see Topic C1). In a typical negative priming study, a given trial requires participants to attend to one stimulus and ignore another stimulus presented at the same time. If the ignored stimulus becomes the target stimulus on the next trial, participants respond more slowly than when a new stimulus is used (see *Fig. 2*). This slowing occurs even if the ignored stimulus moves to a new location, showing that attention is object-based. However, processing of a new stimulus appearing in the location of the previously ignored stimulus is also impaired, showing that attention is also location-based.

Note that there are alternative explanations of the negative priming phenomenon that do not directly invoke the concepts of attention and inhibition. For example, negative priming may reflect the operation of episodic memory. When a to-be-ignored stimulus is first perceived, it is encoded in memory along with a 'do not respond' tag. Presenting that stimulus as a target on a subsequent trial triggers retrieval from memory of the previous encounter, including the 'do not respond' tag. Participants respond more slowly, that is they show negative priming, because of the conflict between the current goal of responding to the stimulus and the remembered goal of ignoring it.

Unilateral visual neglect

Damage to parietal brain areas can cause a condition known as unilateral visual neglect, in which patients appear to ignore stimuli on the side contralateral to the region of brain damage. Typically, damage to the right hand side of the

Trial: 1   2   3

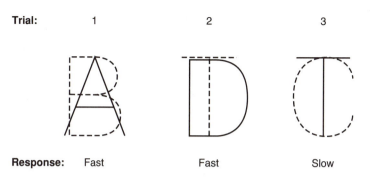

Response: Fast   Fast   Slow

Fig. 2.   Negative priming: the task is to respond as quickly as possible to the figures with a solid outline, and to ignore those with a dotted outline.

brain causes neglect of the left-hand side of space. *Figure 3* shows a neglect patient's attempt at a) a cancellation task and b) a copying task.

One explanation of neglect is that patients have difficulty focusing attention on the neglected side of space. It is not simply a matter of not being able to see that side of space, because strong cueing (e.g. loudly banging the left-hand side of the table at which the patient sits) can attract attention and improve performance on the previously neglected side. Neglect tends to be exacerbated when target stimuli are presented on the 'good' side at the same time as the neglected side, suggesting that patients may also have difficulty disengaging attention from their 'good' side.

Although most accounts of neglect assume that it is a disorder of attention, Bisiach has argued that it may be a disorder of constructing mental representations. The basis for his argument comes from patients' performance when asked to describe from memory the buildings around the Piazza del Duomo in Milan (a familiar scene for them). When asked to imagine that they were standing

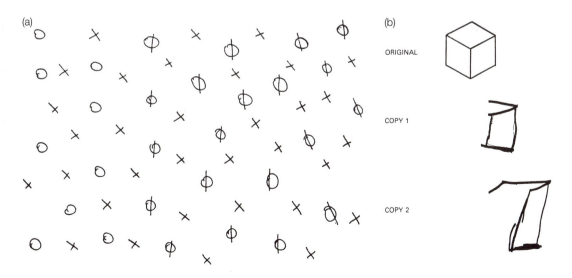

Fig. 3.   Attempts at (a) cancelling Os and (b) copying a square by a patient with left visual neglect (reprinted from A. W. Ellis and A. W. Young (1988), Human Cognitive Neuropsychology. *Psychology Press, Hove, U.K. with permission from Psychology Press).*

facing the cathedral, patients failed to report the buildings on the left-hand side of imaginary space. When asked to imagine turning around to face away from the cathedral, patients accurately described the previously ignored buildings and now omitted information about the buildings they had correctly described before.

Tipper and Behrmann (1996) asked patients with neglect to respond to targets appearing on either side of a computer screen or on either side of a dumbbell-shaped object portrayed on the screen. As expected, patients failed to detect targets appearing on the left-hand side of the screen or the static object. More surprisingly, they also failed to detect targets on the left-hand side of a slowly rotating object. That is, they neglected targets appearing on what was originally the left-hand arm of the dumbbell even when it had rotated through 180 degrees to the right-hand side of the computer screen. Tipper and Behrmann used this finding to argue that attention is both object- and location-based.

C3 DIVIDED ATTENTION AND DUAL TASK PERFORMANCE

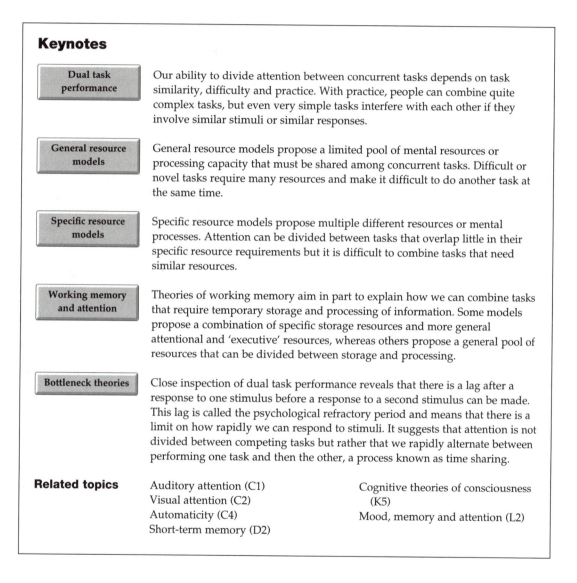

Keynotes

Dual task performance	Our ability to divide attention between concurrent tasks depends on task similarity, difficulty and practice. With practice, people can combine quite complex tasks, but even very simple tasks interfere with each other if they involve similar stimuli or similar responses.
General resource models	General resource models propose a limited pool of mental resources or processing capacity that must be shared among concurrent tasks. Difficult or novel tasks require many resources and make it difficult to do another task at the same time.
Specific resource models	Specific resource models propose multiple different resources or mental processes. Attention can be divided between tasks that overlap little in their specific resource requirements but it is difficult to combine tasks that need similar resources.
Working memory and attention	Theories of working memory aim in part to explain how we can combine tasks that require temporary storage and processing of information. Some models propose a combination of specific storage resources and more general attentional and 'executive' resources, whereas others propose a general pool of resources that can be divided between storage and processing.
Bottleneck theories	Close inspection of dual task performance reveals that there is a lag after a response to one stimulus before a response to a second stimulus can be made. This lag is called the psychological refractory period and means that there is a limit on how rapidly we can respond to stimuli. It suggests that attention is not divided between competing tasks but rather that we rapidly alternate between performing one task and then the other, a process known as time sharing.

Related topics	Auditory attention (C1)	Cognitive theories of consciousness
	Visual attention (C2)	(K5)
	Automaticity (C4)	Mood, memory and attention (L2)
	Short-term memory (D2)	

Dual task performance	Research into divided attention addresses the question of which factors determine our ability to do two things at once. Sometimes it is relatively easy to perform two tasks simultaneously, for example eating while watching television, but other tasks are hard to combine, for example using a mobile phone has been shown to impair driving performance (see Topic A2). Factors such as different stimulus modalities, that make it easy to ignore distractors when

selective attention is required, generally help us to divide attention between tasks.

Task similarity makes dual task performance more difficult. Allport, Antonis and Reynolds (1972) asked participants to shadow an auditory message and then measured their memory for words in a message presented to the other ear. Memory for the words in the non-shadowed message was poor. However, when the non-shadowed words were presented visually so participants read them rather than heard them, memory for them was better. Memory for the non-shadowed stimuli was best of all when those stimuli were pictures rather than words (see Topic C1).

Research into **working memory** (see Topic D2) has used dual task procedures extensively to tease apart the cognitive processes underlying performance of complex tasks thought to require working memory. The pattern of findings is generally that:

- Relatively low-level tasks, particularly those requiring short-term storage of stimuli, interfere with each other if they require processing in the same modality. For example, two verbal short-term memory (STM) tasks, such as remembering digits and remembering letters, interfere substantially with each other. Even a very simple concurrent verbal task, like saying 'the' over and over, disrupts verbal short-term memory (STM). Low-level tasks in different modalities can be combined much more easily, thus concurrent articulation has little effect on visual or spatial STM and it is possible to perform verbal STM and visuo–spatial STM tasks concurrently with little detriment to performance on either task.

- Higher-level tasks, particularly those such as random number generation and verbal fluency that tap central executive resources of planning or retrieval from long-term memory, are difficult to combine.

- There is a small detrimental effect of combining any two tasks, even if they require processing in different modalities. This 'dual task effect' is usually attributed to the central executive loads imposed by the two tasks.

Task difficulty also affects dual task performance, but it is hard to define exactly what makes one task difficult and another easy. Generally, a difficult task is one that is novel and unpracticed, taps multiple cognitive processes, involves rapidly changing sensory input, and involves an inconsistent or illogical mapping between stimuli and appropriate responses to them.

Practice makes tasks easier in themselves and easier to combine with other tasks. Practice at doing two things at once is particularly helpful for improving dual task performance. Spelke, Hirst and Neisser (1976) trained participants to read stories and write words to dictation simultaneously. Initially, these tasks were very hard to combine and participants missed some of the dictated words and read very slowly. After weeks of training, however, their reading speed and comprehension was as good when combined with the dictation task as when they were doing the reading task on its own. They were also able to recall some of the dictated words, although subsequent research by the same authors showed that performance on the dictation task combined with reading was poorer than when dictation was performed alone. Nonetheless, this is an impressive demonstration of our ability to do two things at once.

General resource models

General resource models of divided attention propose that there is a general pool of cognitive resources that must be shared between tasks. Resources are viewed as a type of 'mental energy' or as the 'mental capacity' of a central processor. Any mental process will require a certain amount of resources or capacity. The more resources that are required by one task, the fewer will be available for performing another task at the same time.

Kahneman's (1973) **single capacity model** is an example of a general resource model. Kahneman assumed a single pool of resources that can be shared among competing tasks. The amount of resources available depends on factors such as arousal and differs across individuals. We have conscious control over how we allocate capacity to different tasks. If the available capacity exceeds the combined demands of the concurrent tasks, there should be no detriment to performance. Performance on either task only suffers when there is competition because the pool of resources is small or both tasks have high resource demands. Payne *et al.* (1994) showed that increasing the difficulty of an auditory task, by reducing the sound quality, increased the extent to which it impaired performance on a concurrent visual task.

Johnston and Heinz (1978) proposed a model of selective attention (see Topic C1) that also explains limitations on divided attention. They assumed that there is a limited capacity pool of resources. If one task requires many resources, then stimuli related to other tasks will be filtered out at an early stage of processing and it will not be possible to divide attention between the demanding task and other tasks. However, if the primary task requires few resources, selection will happen at a later stage and attention can be divided between tasks.

General resource models account for the effects of task difficulty on dual task performance by assuming that difficult tasks need more resources. They account for the effects of practice by assuming that practice reduces the resource demands of a task by increasing the efficiency of task performance (akin to the way practice at running reduces the amount of energy we need to run a certain distance). General resource models cannot easily explain the effects of task similarity on dual task performance.

Specific resource models

Specific resource models (e.g. Navon and Gopher, 1979) propose that there are different resources for different tasks. Resources are viewed as different cognitive processes, some of which may be specific to a particular type of task and others of which may be common to many tasks. Imagine that for some reason you have to move your finger through a printed maze while listening to a string of digits and counting each instance of the number '1'. Beyond the need to initiate both tasks, there is little overlap in the resources needed for performing each task. Only the digit task requires speech parsing and verbal working memory, while only the maze task requires visuo–spatial working memory. However, if you need to tap the table every time you hear '1' in the digit list, then the two tasks will be harder to combine because both will be competing for the resources needed to initiate manual response schemas. Working out the extent to which two tasks will interfere requires a **task analysis** that defines the processes or resources needed by each task. Similar tasks interfere because they require similar processes. Difficult tasks interfere because they require many processes, hence their requirements are more likely to overlap. Practice reduces interference by reducing the extent to which high-level processes such as error detection and strategy selection are needed.

Working memory and attention

Baddeley's working memory model (see Topic D2) proposes a combination of general and specific resources. Modality-specific (verbal or visuo–spatial) temporary storage modules are controlled by a 'central executive', which is a limited capacity pool of general processing resources. However, the central executive component of the model is perhaps the most contentious. One of the problems is that closer investigation of general executive functions suggests a range of different and possibly unrelated processes such as selective attention, task switching, retrieval from long-term memory, and so on. In other words, there is rather little evidence for a general resource model of executive function.

The working memory approach typical of North American researchers makes somewhat different claims about resources. Rather than the separate storage and processing modules seen in Baddeley's model, it assumes a general resource that can be devoted to processing or storage. A difficult processing task (e.g. comprehending difficult sentences) will leave few resources for storage (e.g. remembering those sentences), and vice versa. Tasks that make few demands on either storage or processing capacity are easier to combine than those that make larger demands.

Bottleneck theories

Most of the dual task studies discussed so far show an impressive ability to combine tasks, providing the tasks are not too similar or too difficult. However, a more detailed examination of task combination shows that there is always a cost to doing two things at once or in very close succession. In a typical study, participants are asked to respond to two stimuli, making a different response to each stimulus. If the second stimulus appears very soon (say, 100 msec) after the first, there is a delay in responding that suggests a 'hangover' from processing the previous stimulus. This delay is called the **psychological refractory period**. Welford (1952) argued that the psychological refractory period reflects a processing bottleneck that makes it impossible to select or initiate two responses simultaneously. Further evidence for a unitary **response selection bottleneck** comes from research by Pashler (e.g. 1990), showing that the psychological refractory period persists even when stimuli and responses are very different (e.g. a spoken response and a key press) and when participants have had extensive practice at the two tasks.

Studies of the psychological refractory period suggest that our apparent ability to do two things at once may reflect our ability to switch attention or other resources rapidly between tasks, rather than a genuine ability to divide attention. Rapid switching between tasks is called **time sharing**.

C4 AUTOMATICITY

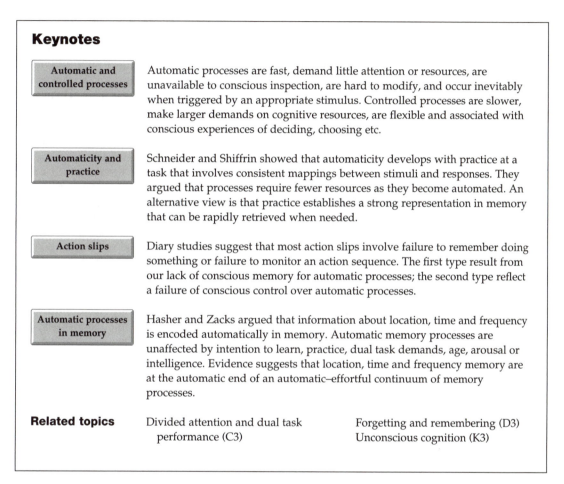

Keynotes

Automatic and controlled processes	Automatic processes are fast, demand little attention or resources, are unavailable to conscious inspection, are hard to modify, and occur inevitably when triggered by an appropriate stimulus. Controlled processes are slower, make larger demands on cognitive resources, are flexible and associated with conscious experiences of deciding, choosing etc.
Automaticity and practice	Schneider and Shiffrin showed that automaticity develops with practice at a task that involves consistent mappings between stimuli and responses. They argued that processes require fewer resources as they become automated. An alternative view is that practice establishes a strong representation in memory that can be rapidly retrieved when needed.
Action slips	Diary studies suggest that most action slips involve failure to remember doing something or failure to monitor an action sequence. The first type result from our lack of conscious memory for automatic processes; the second type reflect a failure of conscious control over automatic processes.
Automatic processes in memory	Hasher and Zacks argued that information about location, time and frequency is encoded automatically in memory. Automatic memory processes are unaffected by intention to learn, practice, dual task demands, age, arousal or intelligence. Evidence suggests that location, time and frequency memory are at the automatic end of an automatic–effortful continuum of memory processes.
Related topics	Divided attention and dual task performance (C3) Forgetting and remembering (D3) Unconscious cognition (K3)

Automatic and controlled processes

The previous topic showed how tasks that were relatively easy and well prac-ticed were easier to perform concurrently than tasks that were difficult and novel. Attention can be divided between two tasks if neither task requires much attention. Another way of describing this situation is to say that two automatic tasks can be combined, providing they involve different sensory inputs and responses, but two controlled tasks will be harder to combine. The distinction between automatic and controlled processing is a common one and has been used to explain a range of findings, from action slips and conscious control of behavior (see Topic K5) to differences in memory for different types of informa-tion.

Schneider and Shiffrin (1977) defined automatic and controlled processes in terms of memory. They assumed that long-term memory (LTM) is a set of inter-connected nodes (see Topic E4) whereas short-term memory (STM) is the subset of those nodes that are currently activated (see Topic D2). They defined an automatic process as the activation of an established sequence of nodes in LTM. A controlled process is the temporary activation, by attention, of a novel sequence of nodes. Thus automatic processes are functions of LTM whereas controlled processes also require STM.

Automatic processes are fast and efficient, making minimal demands on cognitive resources and leaving plenty of processing capacity available for performing other tasks. We are not conscious of performing automatic processes and tend to be unable to recall performing them afterwards. For example, many experienced car drivers report having driven a familiar route from A to B, only to arrive at B and wonder how they got there without an accident because they have no recollection of being 'in control' of their driving. Clearly they were not unconscious while driving, otherwise they would have crashed. But they were driving on auto-pilot using well practiced **action schemas** that required little conscious control and left little impact on memory. Action schemas are action sequences or routines that have been used often and recorded in memory. The drawback of automatic processes is that they are triggered whenever we encounter an appropriate stimulus and, once initiated, they are hard to stop or change. This makes automatic behaviors inflexible.

Controlled processes are slow and attention-demanding. They are costly in terms of cognitive resources but enable flexible behavior. We use controlled processes when responding to novel situations or overriding habitual responses to familiar situations, and also for detecting errors in automatic processes. Controlled processes are associated with experiences of deliberate action, decision making and choosing; they result in explicit memory (see Section D) for the stimuli and actions they encompass. Learner drivers are unlikely to experience the 'driving on auto-pilot' feeling described above because they have to concentrate on every aspect of driving. For them, each component of driving, from changing gear to checking the mirror, requires controlled processing. An experienced driver who changes to a new model of car with a different layout of controls will have to use a controlled processing mode of driving to avoid turning on the wipers each time they want to indicate.

The **Stroop effect** (Stroop, 1935) illustrates the interaction between automatic and controlled processes. The basic Stroop task involves color words (red, yellow, blue, green) printed in different colored inks. The task is to name the ink color of each word as quickly as possible. When the ink color is consistent with the word (i.e. RED is printed in red ink), responses are fast. However, when the ink color is inconsistent (RED is printed in yellow ink), color naming is slow and error-prone. We must use controlled processing to override the automatic process of word reading. Reading is fast because it is automatic, therefore it is difficult to inhibit responses to the words. If the task is changed to one of reading the words, there is little effect of the different ink colors, showing that automatic processes are resistant to interference (see Topic L2 for a version of the Stroop effect using emotional words).

Automaticity and practice

Schneider and Shiffrin (1977) used a visual search task to contrast automatic and controlled processes, and to investigate the development of automaticity through practice. Their basic procedure was as follows: At the start of each trial, participants are shown the target items they will have to search for (e.g. B and Z) and memorize them. Then they see a series of displays containing different items, which they must search for the targets. The participant's task is to decide if the targets appeared in any of the displays. Schneider and Shiffrin manipulated i) the number of target items that had to be memorized and searched for; ii) the number of items in the displays; and iii) the sets of items from which the targets and distractors were drawn. In the **consistent mapping condition**, the target items on every trial came from the same set of items (e.g. numbers) and the distractor items came from a different set of items (e.g. letters). In the **varied mapping condition**, the targets and distractors came from the same set, so a target on one trial could be a distractor on another.

Overall, people were slower in the varied mapping condition. However, the most interesting aspect of the data is the effect of target number and distractor number on performance. In the varied mapping condition, performance got worse (slower with more errors) as the number of targets increased and as the size of the displays increased. In contrast, performance in the consistent mapping condition was relatively unaffected by having to remember more targets or search larger displays (see *Fig. 1*).

Schneider and Shiffrin argued that the different results for the different mapping conditions reflect the operation of controlled and automatic processes. In the varied mapping condition, participants perform a controlled, serial search of each display. This search is harder when there are more targets to look out for and when there are more items in the display. In the consistent mapping

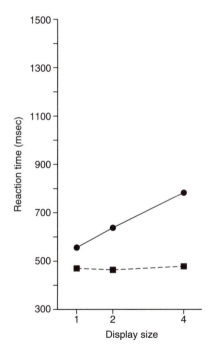

Fig. 1.   *The effect of display size on search times for four target items (positive trials only). When the targets are numbers and the distractors are letters on each trial, searching the displays is a rapid, parallel process unaffected by display size (consistent mapping condition, ■–––■). When the targets and distractors are drawn from the same set of items, say letters, the displays are searched serially and the search gets slower as the display size increases because performance cannot be automated (varied mapping condition, ●——●) (from Schneider, W. & Shiffrin, R. M. (1977) Controlled and automatic human information processing: 1. Detection, search, and attention.* Psychological Review, 84, *1–66. Copyright © 1977 by the American Psychological Association. Adapted with permission).*

condition, participants can perform a rapid search which, because items are searched in parallel, is unaffected by the number of items. Participants can do this because the targets are always numbers and the distractors are always letters (or vice versa) and they have had years of practice at discriminating numbers from letters.

Schneider and Shiffrin further investigated the effects of practice by using one set of letters as the targets and another set of letters as the distractors. Performance in this version of the consistent mapping condition improved considerably with practice, suggesting that participants' search processes became automatic as they learned the target and distractor sets. When the mapping was reversed, so the target letters became the distractors and vice versa, performance was even worse than it had been at the start of the experiment. This dramatic decline in performance shows the difficulty of altering automatic processes once they are established.

Schneider and Shiffrin argued that automatic processes require few resources and are therefore unaffected by capacity limitations in the cognitive system. Thus practice serves to speed up performance of a task by reducing the resources needed. However, practice may also change the way we perform a task, for example it may alter the strategies we use. Logan (1988) proposed an alternative to Schneider and Shiffrin's view, known as **instance theory**. He argued that each trial of responding to a particular stimulus leaves a separate trace in memory, that is, the trial is stored as an instance in memory. With practice, the memory trace becomes stronger and easier to retrieve. Logan proposed that there are two ways of performing a task. You can work out what to do from scratch or you can retrieve the solution from memory. With novel tasks, the first way is the only option. However, with practiced tasks, both ways are viable and the memory retrieval route is fastest. Logan's approach emphasizes the importance of **knowledge** for successful task performance. Novices are slow because they have little knowledge about the task, therefore they must use effortful, controlled processes to reach a solution. Experts are fast because they can automatically retrieve the appropriate solution to a problem from memory (see Topic F3).

Action slips

Reason (1979) asked people to record their action slips (i.e. unintended actions or failures to act) in diaries. Sixty percent of the reported action slips fell into one of two categories:

- **Storage failures** were slips resulting from forgetting actions that had already been performed. For example, one diarist forgot they had just made a pot of tea and began pouring water into the full teapot. Storage failures arise when carrying out habitual action schemas. These routines are performed automatically and, because they are unattended, leave little or no memory trace. This is one reason why it is difficult to remember whether you have already performed an action that you perform every day, such as locking the front door when you leave the house, or switching off the gas.
- **Test failures** were slips resulting from a failure to monitor the progress of an action. Typically, people began performing a task and then lapsed into unintended, habitual behavior. Examples of test failures are picking up a toothbrush to put it away and accidentally starting to brush your teeth with it, or intending to make a cup of black coffee for a friend and adding milk to it because you usually drink white coffee. This type of slip arises because

automated processes happen inevitably when an appropriate stimulus is present and, once begun, overriding them requires conscious attentional control. If cognitive resources are directed elsewhere (e.g. you are talking to your friend while making coffee), there may be insufficient resources left for monitoring performance.

Norman and Shallice (1986) proposed a three-level theory of action control to explain such findings (see also Topic K5). When a stimulus triggers only one particular action schema or response, that action is performed automatically. When a stimulus triggers several action schemas, the strongest or most activated schema inhibits the competing schemas in a semi-automatic response selection process called **contention scheduling**. Contention scheduling is overseen by a control process called the **supervisory attentional system** (SAS). When a habitual action is triggered, but inappropriate, the SAS increases the activation of a more appropriate competing action schema so that action 'wins the competition' to be performed. Thus the SAS can override contention scheduling when we need to make novel responses to familiar stimuli or stop a habitual action routine part way through.

Automatic processes in memory

Hasher and Zacks (1979) used Schneider and Shiffrin's distinction between controlled and automatic processing to explain the differences in memory for different types of information. They argued that whereas some processes become automatic through practice, others are innately automatic. Included among these innate automatic processes are those that encode information about **context**, that is, spatial location, timing and frequency, in memory. Most of us have relatively poor memory for these features of events, yet the fact that we remember them at all is remarkable considering we usually make no effort to do so. For example, you may be able to remember where in this book you saw an illustration of the Müller–Lyer illusion, and even which side of the page it is printed on. You may be able to remember whether you read the introductory chapter before the other chapters, and how often you read each chapter. Hopefully though, you were concentrating on reading the chapters and understanding the information in them, rather than trying to memorize this peripheral information.

Hasher and Zacks listed six criteria for automaticity. Automatic memory processes are unaffected by:

(i) **Intention to learn**. Context information is remembered moderately well under **incidental learning** conditions, where the participant is unaware of the impending memory test and makes no effort to learn, and is not remembered any better when there is a deliberate effort to learn.

(ii) **Practice**. Location, time and frequency are already encoded by automatic processes so there can be no benefit of practice.

(iii) **Concurrent task demands**. Automatic processes exert minimal loads on cognitive resources, therefore they do not interfere with other tasks performed at the same time and are not themselves impeded by those tasks.

(iv) **Age**. Automatic processes develop early and then do not improve with age. They are already efficient, so increases in resource capacity or performance efficiency with age do not benefit memory for automatically encoded information.

(v) **Arousal**. Variations in stress or arousal levels, that affect the availability of mental resources, do not affect automatic processes because those processes are only minimally dependent on those resources.

(vi) **Individual differences**. Automatic processes are not affected by individual differences in intelligence that might affect encoding strategy, resource availability etc.

Tests of Hasher and Zacks' framework for classifying memory processes have investigated the effects of these six variables on memory for context. For example, Andrade and Meudell (1993) asked participants to count aloud while words appeared in different locations on a computer screen. Counting by sevens impaired memory for the words, compared with counting by ones, but had no effect on memory for the locations in which those words were presented (see *Fig. 2*). Although there is debate about whether location, time and frequency memory are purely automatic (some researchers have found effects of the above variables), the evidence suggests that they are at the automatic end of an automatic–effortful continuum.

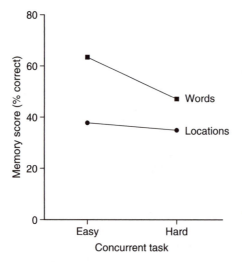

Fig. 2. The effect of counting by ones (easy task) versus counting by sevens (hard task) on memory for words and their spatial locations (Andrade and Meudell, 1993).

D1 INTRODUCTION TO MEMORY

Keynotes

What is memory?

Memory comprises the processes that encode, store and retrieve information. Short-term memory (STM) is a transient store of recently encountered information whereas long-term memory (LTM) is more permanent. LTM can be subdivided into episodic or event memory (which includes autobiographical memory), semantic memory and prospective memory. It can also be divided into declarative or reportable memory, and procedural or skill memory. A recent distinction is between conscious or explicit memory and unconscious or implicit memory.

How can memory be studied?

There is a tension between studying memory in idealized conditions and studying it in real-life contexts that give noisier but more ecologically valid results. Ebbinghaus took the former approach, testing his memory for nonsense syllables in tightly controlled conditions. He discovered that memory improves as a linear function of the time spent learning, that distributing learning over several periods results in better memory than trying to learn everything in one go, and that memory varies according to the serial position of items in the learned list. Bartlett took the latter approach, studying memory for complex stimuli such as stories and pictures. He argued that memory is an active process involving understanding, interpretation and reconstruction.

Related topics   Methods of investigation (A2)   Short-term memory (D2)

What is memory?

Memory refers to our ability to retain information about past experiences. It encompasses the processes by which we acquire, record or **encode** information, the processes by which we **store** information in an accessible format, and the processes by which we can later **retrieve** that information.

Memory can be classified in different ways. In 1890, William James described two types of memory: **primary memory**, which is memory for events that have just happened; and **secondary memory**, which is memory for events that happened some time ago. Thus primary memory is a relatively transient store, whereas secondary memory is long-lasting. Primary memory and secondary memory are now more commonly referred to as **short-term memory (STM)** and **long-term memory (LTM)** respectively. LTM can be subdivided into memory for specific events or episodes that we have experienced, or **episodic memory**, and memory for general information or facts, called **semantic memory**. Semantic memory includes our knowledge of word meanings and therefore is an essential basis of language. Semantic and episodic memory are types of **retrospective memory**; memory for events and information encountered in the past. **Prospective memory** is remembering to do things in the future, such as attending a tutorial on Friday or buying milk on the way home today. Another way of classifying LTM is to divide it into **declarative memory**, that is, those memories that we can report – such as our knowledge that Paris is the capital of France and our memory of a holiday – and **procedural memory**, which includes skills like riding a bike or kicking a football. Since the 1980s, a distinction has also been made between **explicit** and **implicit** memory. Explicit memory refers to those memories that we are aware of and can report. It is measured using tests of **recall** or **recognition**. Implicit memory refers to unconscious memories that are revealed when a prior experience brings about changes in performance in the absence of conscious or explicit recollection of that experience. For example, hearing the word 'javelin' during an experiment may increase your tendency to respond with 'javelin' when asked to name a word beginning 'ja–'. This is an example of a **word stem completion** test (see Topic D3).

Memory is important because it underpins almost every aspect of our behavior and personalities. We can perceive stimuli more easily if we have a memory of encountering them before. Developing semantic memory, for example learning our native language, is an essential part of childhood. Our ability to make friends depends on our face recognition memory and knowledge of the rules of social interactions such as conversations. Our individual personalities are influenced by our **autobiographical memory**, memory for our personal history (see Topic D4).

How can memory be studied?	Two pioneers of memory research, Hermann **Ebbinghaus** (1850–1909) and Sir Frederick **Bartlett** (1886–1969), took very different approaches to studying memory. Both approaches continue to influence memory research. Ebbinghaus pioneered techniques for studying memory in idealized conditions, as far as possible avoiding variations in the learning environment and individual differences in background knowledge. He tested himself repeatedly, using highly controlled testing conditions (e.g. testing himself at the same time each day) and avoiding the influence of prior knowledge by testing memory for **nonsense syllables**, meaningless consonant–vowel–consonant trigrams such as ZOD, PIV. He quantified learning using the **method of savings**, whereby he learned a list of nonsense syllables on one occasion and then measured how many learning trials he needed to relearn the list on another occasion. In one experiment, Ebbinghaus investigated the effect of practice on learning. He learned lists of 16 nonsense syllables, repeating them 8, 16, 24, 32, 42, 53 or 64 times on day 1 and then recording how many learning trials he needed on day 2 to be able to recite them perfectly. Two important findings emerged from this experiment: i) the amount learned increases linearly with the time spent learning (the **total time hypothesis**); and ii) **distributed practice** is better than massed practice, that is, it was better to spread the learning over 2 days than to try and do it all in one go (*Fig. 1*). Ebbinghaus also described the **serial position curve** in free recall (*Fig. 2*), that is the tendency for items near the start of a list to be remembered well (the **primacy effect**) and for those near the end of a list to be remembered well (the **recency effect**) relative to the less well remembered items in the middle of a list. The advantage of Ebbinghaus's approach is that it reveals laws and functions that underpin all memory function, but which might be obscured by environmental noise and individual differences if memory were examined in less tightly controlled conditions.

Bartlett's approach emphasized the complexity of memory in the real world. He argued that memory is an active process and tested it by asking people to

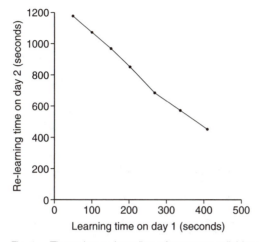

Fig. 1.   *Time taken to learn lists of nonsense syllables on day 2 as a function of time spent learning them on day 1 (redrawn from Ebbinghaus, 1885). When the learning was spread evenly across the two days (with 64 trials or around 400 seconds on day 1), the total time needed to learn the list across the two days was only about two thirds of the time needed when the minimum time was spent learning on day 1 (8 trials or around 50 seconds). This is a demonstration of the advantage of distributed practice over massed practice.*

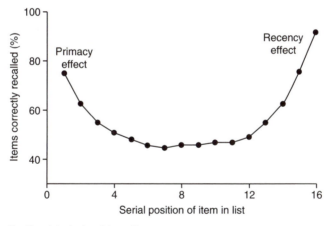

Fig. 2.   *A typical serial position curve.*

recall meaningful but complex stimuli such as stories and pictures. Often these stimuli were unfamiliar to the people Bartlett tested, for example the native American folktale called the War of the Ghosts. By studying the errors people made when recalling the stories or reproducing the pictures (*Fig. 3*), Bartlett drew two conclusions about human memory. It involves **effort after meaning**, that is people try to make sense of new information by relating it to their existing knowledge, and it involves **reconstruction**, that is people use their existing knowledge to fill in details in their memory. The advantage of Bartlett's approach is that it has **ecological validity**, it tells us how memory works in the real world (see Topic A2).

Early findings from both research traditions are still highly relevant. For example, when learning psychology, you should devote sufficient time to

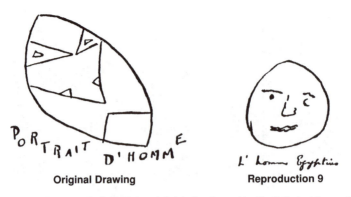

Original Drawing   **Reproduction 9**

*Fig. 3. An example (left) of the pictorial stimuli used by Bartlett, and the end result (right) of
'serial reproduction', in which the first participant draws the figure from memory and passes
their drawing to the next participant, who draws that from memory, and so on (reproduced
with permission from Batlett, F.C. (1932)* Remembering, *Cambridge University Press,
Cambridge).*

revision but, within that time, revise a particular topic in several short periods
rather than all at once. You may need to make more effort to learn information
from the middle of a course than information in the first and last lectures of
that course. The most efficient way to learn is to work out how the new information fits with what you already know. You do not need to learn everything
word-perfectly. If you have a good understanding of the material, you will
be able to reconstruct some of the information rather than having to recall it
verbatim.

D2 SHORT-TERM MEMORY

Keynotes

The structure of memory

Information from our senses enters short-term memory (STM) via sensory registers. STM is a temporary store for relatively unprocessed (sound-based or visual) information. Verbal STM has a limited capacity of 7 ± 2 items, which can be used most efficiently by chunking the information to be remembered. LTM stores a practically limitless amount of information in meaning-based format. Evidence for the STM–LTM distinction came from recency effects in free recall and double dissociations in the effects of brain damage on memory. However, recency is now considered a general feature of memory rather than evidence for a separate STM. Patients with damaged STM and intact LTM suggest that STM is a multifaceted system.

Working memory

Working memory refers to the processes that temporarily store and manipulate information so that we can use that information to follow a conversation, solve problems etc. Baddeley and Hitch's model of working memory comprises the phonological loop (PL), visuo-spatial sketchpad (VSSP), central executive and, recently, the episodic buffer. These subsystems have limited capacity. The PL and VSSP store modality-specific (auditory, visual) information, supporting language learning and mental imagery functions, whereas the episodic buffer stores multimodal information. The central executive processes and uses this information, serving functions of selective attention, strategy selection, LTM retrieval, and co-ordination of concurrent tasks. Individual differences in working memory capacity are associated with differences in performance of complex cognitive tasks such as intelligence tests and reading comprehension.

Measurement of short-term memory

Digit span, matrix span and the Corsi blocks task measure verbal, visual and spatial STM performance respectively. The Wisconsin card sorting task, verbal fluency and random number generation provide measures of central executive function. Working memory span measures the overall efficiency of working memory as the ability to store and process information simultaneously.

Quantitative models of short-term memory

Computational or mathematical models that attempt to mimic human performance on immediate serial recall tasks suggest three possible mechanisms underlying verbal STM: (i) items are stored alongside information about their position in the list, and positional context cues recall; (ii) items are linked to other items in the list, and recall of previous items cues recall of the current item; or (iii) the activation of an item's representation in memory varies according to list position and the most active items are recalled first.

Related topics

Divided attention and dual task performance (C3)

Introduction to memory (D1)

Forgetting and remembering (D3)

Cognitive theories of consciousness (K5)

Mood, memory and attention (L2)

The structure of memory

Several influential models of memory in the 1950s and 1960s built on the distinction made by William James between a temporary short-term memory (STM) and a more permanent long-term memory (LTM). The model proposed by Atkinson and Shiffrin in 1968 was typical of the models around this time and therefore was termed the **modal** ('most common') **model** of memory (*Fig. 1*). It is a **multi-store model** of memory. They suggested a tripartite memory structure. Incoming information from the environment is very briefly captured in **sensory registers.** Some of the information in the sensory registers is then transferred to the **short-term store**, from which it may be lost through decay or transferred (copied) via **rehearsal** to the **long-term store**. Information in the long-term store may be lost through decay or interference, or may be output via the short-term store. The short-term store has a limited capacity of seven plus or minus two items (Miller, 1956). The size of the items is variable, thus this nine-letter list is hard to recall – ZHKXMSJYP – whereas this nine-letter list is easy to recall – CATDOGFOX. **Chunking** the second list into words aids recall, effectively increasing the capacity of STM (see also Topic D3). LTM has a practically unlimited capacity. Baddeley (1966a,b) showed that stimuli that sound alike (MAN, CAN, CAT, MAP, CAP) are hard to recall from STM whereas stimuli with similar meanings (LARGE, BIG, HUGE, GRAND) are hard to recall from LTM. This finding suggests that STM stores information in a sound-based form whereas LTM stores it in a semantic or meaning-based form.

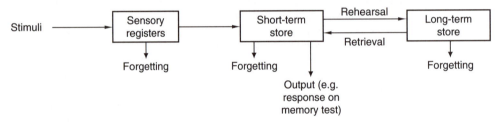

Fig. 1.   The tripartite model of memory popular in the 1960s

Evidence for a visual sensory register, or **iconic store**, comes from research by Sperling (1960) using a **partial report technique**. A 3 × 4 array of letters is presented very briefly. Participants are typically unable to report more than a few of the letters before their image of the array decays. However, if presentation is followed immediately by a cue to report just one particular row of the array then recall is accurate. This research suggests that accurate visual sensory images last for a few hundred milliseconds. The equivalent register for auditory information is termed the **echoic store** (see also Topic B1).

One source of evidence for two separate memory systems, STM and LTM, is the typical serial position curve obtained in free recall tasks (see *Fig. 2* in Topic D1). Atkinson and Shiffrin argued that the last items in a long list can be recalled easily because they are still in STM, whereas early and middle items must be retrieved from LTM before being output from the STM. Different variables affect recall of items from the end of the list and recall of earlier items. Recall of the last few items is disrupted by interposing additional items or another task in the retention interval between stimulus presentation and recall. The additional items, or additional information involved in the interpolated task, fill up the 'slots' in STM, displacing items from the original list which consequently lose their recall advantage. Recall of early and middle list items, but not recall of the last few items, is influenced by stimuli that have general effects on cognitive function, for example age and concurrent task demands. The primacy effect is attributed to increased rehearsal of the first items in a list or to their increased distinctiveness by virtue of being at the start of the list. It does not represent output from a separate memory system.

Neuropsychology provides another source of evidence for two memory systems, showing a **double dissociation** (see Topic A4) between damage to LTM function and damage to STM function (see also Topic D4). Milner (1966) reported a case study of a patient known as H.M. who performed normally on tests of STM but could not recall or recognize stimuli after a short delay, that is, his episodic LTM was impaired. H.M. seemed to have lost the ability to transfer information into LTM. For example, every time he was told that his uncle had died, he responded with the same intense grief as the first time he heard that news. He compared his experience, and the problem of keeping track of what has just happened, to waking from a dream. Warrington and Shallice (1969) reported a patient with the converse pattern of impairment. K.F. had good recall of stories and long lists of words but impaired STM performance.

Problems for the modal model of memory arose from these same two sources of evidence. Recency effects have been observed even when the information recalled must be in LTM, for example, details of a season's rugby matches. Thus recency may be a general feature of memory rather than a specific feature of STM. It has been attributed to the increased distinctiveness of recent memories relative to older memories. Just as when you walk along a busy street, people just in front of you appear as individuals whereas those further away appear as a mass of almost indistinguishable heads, so recent memories are easier to distinguish, and thus easier to recall, because of their proximity to the present time. K.F.'s pattern of impairment also raised problems for Atkinson and Shiffrin's view. According to the modal model, information enters and leaves LTM via STM. K.F.'s damaged STM should have caused problems for his LTM but it did not, suggesting that there are routes in and out of LTM that are independent of STM. A third problem for Atkinson and Shiffrin's model came from demonstrations that rehearsal does not always aid encoding of information in LTM. Some types of rehearsal are more effective than others (see Topic D3). Some researchers argued for a single memory system. For example, Norman (1968) argued that identification of familiar stimuli could not be so rapid if incoming perceptual information had to pass through STM before reaching LTM. He suggested that STM phenomena result from the temporary excitation of representations in LTM (see Topic E1 for more information about representations).

Working memory

In 1974, Baddeley and Hitch proposed a new model of STM that shifted the focus from memory structure to memory processes and functions. They argued that STM serves a key role in cognition, temporarily storing information so that it can be used in cognitive tasks such as reasoning. They referred to this function of STM as **working memory**. Their working memory model encompasses temporary storage functions (i.e. STM) and manipulation of the stored information for use in complex cognition (i.e. working memory). They falsified Atkinson and Shiffrin's assumption that a unitary STM served as a working memory by showing that remembering a short list of digits (an STM task) caused relatively little disruption to reasoning, comprehension and LTM tasks.

They suggested that the working memory processes needed for those tasks were separate from those needed for temporarily storing the digits. Reasoning, comprehension and retrieval from LTM were functions of a general processing system termed the **central executive**, whereas short-term storage of verbal stimuli was a function of the **phonological loop** (PL). K.F.'s pattern of memory performance can be explained by assuming that he has a damaged PL but an intact central executive that allows information to flow into and out of LTM. Baddeley and Hitch also proposed a temporary store for visual and spatial information, the **visuo–spatial sketchpad** (VSSP). In 2000, Baddeley proposed a new component in the working memory model, an **episodic buffer** that stores multimodal information and serves as an interface between working memory, LTM and consciousness. Baddeley's revised model is illustrated in *Fig. 2*.

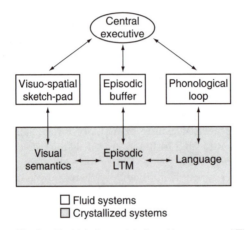

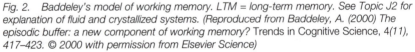

Fig. 2. Baddeley's model of working memory. LTM = long-term memory. See Topic J2 for explanation of fluid and crystallized systems. (Reproduced from Baddeley, A. (2000) The episodic buffer: a new component of working memory? Trends in Cognitive Science, 4(11), *417–423. © 2000 with permission from Elsevier Science)*

The subsystems of working memory have limited capacity, making it difficult to carry out two tasks concurrently if they both require the same subsystem (see Topic C3). If the assumption of separate verbal and visuo–spatial subsystems is correct, then a concurrent verbal task should disrupt verbal STM (by taking up some of the limited capacity of the PL) but have little effect on visual STM. This is generally the case, thus support for the working memory model comes from demonstrations that repeating a simple word (an interference technique known as **articulatory suppression**) impairs verbal STM (Murray, 1967) but not visual STM (Smyth, Pearson and Pendleton, 1988), whereas concurrently creating a visual image or tapping a spatial pattern on a keyboard impairs visual STM more than verbal STM (e.g. Logie, Zucco and Baddeley, 1990). In general, there is a **crossover interaction** between the modality of the STM task and the modality of the competing task (*Fig. 3*).

The PL comprises a storage component, the **phonological store**, and a rehearsal component, the **articulatory loop**. Information is thought to decay from the store in about 2 seconds unless rehearsed. The **phonological similarity effect**, the finding that similar-sounding stimuli like PVDCTG are harder to recall than dissimilar stimuli like KZHMJL (Conrad and Hull, 1964), supports

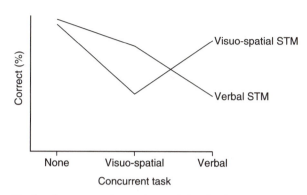

Fig. 3.   A cross-over interaction: verbal tasks disrupt verbal short-term memory (STM) performance more than visuo-spatial tasks do, whereas visuo-spatial tasks disrupt visuo-spatial STM performance more than verbal tasks do. This pattern of data supports the hypothesis of separate visuo-spatial and verbal stores in working memory.

the assumption that the PL stores information in a sound-based form. The **irrelevant speech effect**, the disruptive effect of irrelevant, meaningless background speech on verbal STM, also suggests sound-based storage (Colle and Welsh, 1976). The argument from a working memory perspective is that the incoming speech gains obligatory access to the phonological store and disrupts representations stored there. An alternative view is that the background speech makes it hard for attentional processes to keep the to-be-remembered items in a coherent, temporally ordered stream (see Topic C1). The concept of a sub-vocal rehearsal mechanism is supported by the effect of articulatory suppression, and the finding that words that take longer to say are harder to recall than shorter words (the **word length effect**; Baddeley, Thomson and Buchanan, 1975). Note however that there have been problems replicating the word length effect. When word lists are matched for numbers of phonemes and phonological similarity, those with words that take longer to say because they have longer vowel sounds are remembered as well as those with words that take less time to pronounce (Lovatt, Avons and Masterson, 2000).

Visuo–spatial STM shows a similar range of effects, for example a **visual similarity effect** and an **irrelevant pictures effect**, suggesting that the VSSP may be functionally analogous to the PL. Thus, visually similar stimuli are harder to remember than dissimilar stimuli, even when those stimuli are verbal such as letters or words, hence 'fly, cry, dry' is harder to remember than 'guy, sigh, lie' (Logie *et al.*, 2000). Visual working memory tasks, such as using imagery mnemonics to learn word lists (see Topic D3), are disrupted more than verbal tasks, such as learning by rote rehearsal, by irrelevant visual stimuli such as abstract pictures or a flickering array of black and white dots (Quinn and McConnell, 1996). These findings support the claim of a separate STM store for visuo–spatial information. The concept of a visuo–spatial rehearsal mechanism is supported by the selective effects of **concurrent spatial tasks**, particularly those involving eye movements, on performance of visuo–spatial STM tasks (Pearson and Sahraie, 2003), and the effects of **complexity** (complex, zigzagging routes are harder to recall than simple, linear or circular routes, Kemps, 2001). Logie (1995) refers to the visuo–spatial store as the **'visual cache'** and to the rehearsal mechanism as the **'inner scribe'**.

What are the uses of STM? Baddeley, Gathercole and Papagno (1998) argued that good PL function is important for language learning, enabling novel sounds to be chunked and transferred to LTM. They based their conclusion on evidence that children with good verbal STM learn their native language faster than those with poorer verbal STM (Gathercole and Baddeley, 1989), children and adults with better verbal STM learn second languages more easily (e.g. Service, 1992), and impaired verbal STM is associated with difficulties acquiring language (e.g. Bishop *et al.*, 1996). A counter-argument is that, rather than STM aiding language learning, good language skills aid verbal STM.

The VSSP serves visual imagery functions as well as STM functions (see Topic E2). It supports verbal learning via **imagery mnemonic** strategies (see Topic D3). Visuo–spatial working memory is also important for creativity. Pearson, Logie and Gilhooly (1999) found that a concurrent spatial tapping task, designed to load the VSSP, reduced the number of legitimate patterns generated on a mental synthesis task (see *Fig. 4*). Their finding suggests that we use the VSSP for visualizing novel solutions to problems.

'circle, square, circle, rectangle, figure 8'

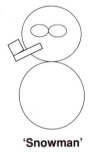

'Snowman'

Fig. 4. Example of a trial on the mental synthesis task used by Pearson, Gilhooly and Logie (1999). Participants are told a selection of elements and asked to mentally combine those elements to form a nameable and recognizable figure.

The central executive acts as a control system, enabling information in the PL and VSSP to be used in complex cognitive tasks. It is thought to serve functions of selective attention, co-ordination of concurrent tasks, retrieval from LTM, inhibition of irrelevant information, strategy selection, and task switching. There is currently debate about whether these processes are functions of a unitary central executive, or whether there are several executives – an 'executive committee' (Baddeley, 1996). In support of the 'unitary central executive' view, dual task studies have shown mutual interference between some of these processes. For example, Baddeley, *et al.* (1998a) used a **random generation** task as one measure of executive function. They assumed that pressing keys in a random order without lapsing into stereotyped or repetitive sequences requires the executive functions of strategy selection and strategy switching. They asked participants to do the random key pressing task at the same time as performing other tasks that load putative executive resources, for example problem solving or verbal fluency tasks (e.g. retrieving as many animal names from LTM as possible in the time available). These secondary tasks impaired performance on the random generation task, making the sequences of key presses less random. Participants found it particularly difficult to combine the key pressing task with a task that required switching between saying the alphabet and counting (to generate responses such as 'C – 5 – D – 6 – E – 7 …'). In support of the 'executive committee' view, correlational studies suggest that there is rather little relationship between different executive functions. Thus Miyake and colleagues (2000) showed that measures of executive functions such as task switching and inhibition of irrelevant information correlated quite modestly with each other and made separate contributions to performance on complex tasks such as random generation.

While European research into working memory has typically focused on the storage functions of working memory (the PL and VSSP), North American researchers have focused more on the processing aspect (which some equate to the central executive), using correlational studies of **individual differences** to explore the relationship between working memory and complex cognitive functions. This research uses **working memory span** (see below) to measure overall working memory capacity. For example, Daneman and Carpenter (1980) showed strong correlations between working memory span and reading ability. Working memory span correlates with measures of intelligence (see Topic J2). It is associated with ability to retrieve information from LTM, so people who score highly on working memory span tasks also tend to be good at verbal fluency tasks (Rosen and Engle, 1997). Working memory span also correlates with ability to ignore irrelevant information. For example, Conway, Cowan and Bunting (2001) showed that people with high working memory span scores were less likely than those with low span scores to hear their name spoken in the distractor message of a **dichotic listening task** (see Topic C1). Engle (2002) argues that the essential component of good performance on working memory tasks is the ability to control attention. It is this ability to attend selectively to the currently important aspects of complex cognitive tasks that explains the relationship between working memory capacity and performance on tests of intelligence, reading comprehension, LTM and so on.

The **episodic buffer** is a limited capacity subsystem of working memory that stores multimodal information and serves as an interface between working memory and LTM. The hypothesis of this additional subsystem aims to explain how information stored in the VSSP and PL can be integrated, and combined with information from LTM, so that unified representations are available for performing cognitive tasks. Although still in its infancy and controversial, the episodic buffer hypothesis potentially helps explain the mutual influences of working memory and LTM, and the unified conscious experience of information (e.g. mental images) held in working memory (Baddeley, 2000).

The **relationship between working memory and LTM** has been thought of in various ways. In Baddeley and Hitch's (1974) model, working memory served as a gateway into LTM, receiving information from sensory processes and transferring that information to LTM. An alternative view is that working memory is the currently activated subset of LTM representations, with attentional processes offsetting the tendency for activation levels to decay over time (e.g. Cowan, 1993; Norman, 1968; Schneider and Shiffrin, 1977 – see Topic C4). Logie (1995) proposed another view, that working memory served as a workspace, separate from LTM, for manipulating activated LTM representations.

Measurement of short-term memory

Verbal STM is often measured by a **digit span** task, requiring participants to listen to a list of digits and repeat the list immediately, retaining the original serial order of the digits. Performance on this task is affected by rehearsal ability. A purer measure of PL storage capacity is provided by Gathercole's non-word repetition test, which requires repetition of increasingly long non-words (see Topic J1). This test is particularly useful in assessing verbal STM in children who have not yet learned to rehearse. **Matrix span** provides a measure of visual STM. Participants view a grid of black and white squares for a few seconds and then recall which squares were black by marking the appropriate squares on a blank grid. The **Corsi blocks test** measures spatial STM. The experimenter taps a sequence of blocks from an array of nine blocks and the participant attempts to replicate the sequence, tapping the same blocks in the same order. There is disagreement about how best to measure central executive function. Some tests are often used in clinical settings, for example, the **Wisconsin card sorting test** (which measures rule learning and strategy switching) and **verbal fluency** (which measures how quickly people can retrieve categorized information, e.g. animal names, from LTM). Other tests are used primarily for research (e.g. **random number generation**, which requires strategy switching and attention to avoid lapsing into stereotyped response sequences, such as 2 4 6 8). The overall efficiency of working memory can be measured using a **working memory span** task that involves simultaneous processing of information (e.g. sentence comprehension or mental arithmetic) and storage of the products of that processing (e.g. the last word in each sentence or the answer to each sum).

Quantitative models of short-term memory

Computational or mathematical models provide a way of testing hypotheses about human memory and help to develop theories of memory by encouraging a better specification of the processes involved (see also Topic A4). If a theory of human memory is implemented on a computer, which fails to perform in a similar way to the human, we assume the theory is incorrect. If the computer successfully mimics human performance, then the theory gives a possible account of human memory.

Quantitative models of short-term memory have focused on immediate serial recall of verbal stimuli (e.g. digit span). Several possible mechanisms of verbal STM have been modeled:

- **Position–item associations**: in models such as that of **Burgess and Hitch (1999)**, each list item is stored as a single connectionist node (see Topic E4) associated with information about its position or context in the list. At recall, reinstating information about position serves to reactivate the associated item node.
- **Item–item chaining**: in models such as the **theory of distributed associative memory**, or **TODAM** (Lewandowsky and Murdock, 1989), order information is represented by linking together ('chaining') representations of the list items in order of presentation. Recall of one item cues recall of the next item in the list. Chaining models tend not to replicate the pattern of errors made by humans on serial recall tasks.
- **Activation strength**: in the **primacy model** (Page and Norris, 1998), the activation strength of list items in memory decreases across the list, forming a 'primacy gradient'. Activation across the list decays with time. At recall, the most active item is recalled first, and its representation is then deactivated so it is unlikely to be recalled again.

D3 FORGETTING AND REMEMBERING

Keynotes

Forgetting	Memory is selective. We lose detailed information rapidly, from sensory memory, short-term memory (STM) and long-term memory (LTM), but remember the gist in LTM indefinitely. Decay theory explains forgetting as a constant rate of fading with time. Interference theory explains forgetting as the disruptive effect of new memories on old (retroactive interference) and of old memories on new (proactive interference). Sleeping immediately after learning may aid memory by reducing retroactive interference. Changing the topic causes release from proactive interference.
Encoding	STM can be improved by chunking information. Simple repetition or maintenance rehearsal has little benefit for LTM. Elaborative rehearsal aids memory by adding meaning, structure, and retrieval cues to the encoded information. Craik and Lockhart argue that memory depends on the level of processing at encoding.
Storage	Collins suggests that semantic memory is organized in a roughly hierarchical fashion. Concepts are stored as nodes, linked to related concepts. At recall, activation spreads through links, reaching related concepts faster than unrelated. In the spreading activation model, typical items are linked to concepts more strongly than atypical, explaining why people are faster to verify 'a robin is a bird' than 'an ostrich is a bird'. Alternative approaches include parallel distributed processing models, where concepts are represented by patterns of activation. Scripts and schemas are storage structures for frequently-used knowledge about complex concepts and situations.
Retrieval	Recognition involves two processes, a judgment of familiarity and a memory decision. Recall may also involve two processes, generation of possible answers followed by recognition of one or more of them. Successful recall is more likely when there is a match between the participant's mood, internal state and external context at encoding and retrieval.
Implicit memory	Implicit memory can be revealed as response behavior changes on tasks that require no deliberate retrieval of memory. Implicit and explicit memory are differentially sensitive to variables such as attentional load.
Related topics	Short-term memory (D2) Concepts (E3) Limits and failures of memory (D4) Connectionism (E4)

Forgetting   Memory is selective; it works efficiently because we forget things. In general, we remember gist and forget verbatim information and low-level perceptual

information. For example, we remember the meaning of a statement but forget
exactly how it was phrased and the voice pitch or print size in which it was
delivered. Typical forgetting curves show that we lose a lot of information very
soon after learning but then retain the remaining information almost indefinitely (see *Fig. 1*). Thus Bahrick (1984) showed that English-speaking participants quickly forgot much of the Spanish they learned at school, but retained
some knowledge of Spanish even 50 years later. Much information loss occurs
from the sensory registers and short-term memory (STM) but some also occurs
from long-term memory (LTM).

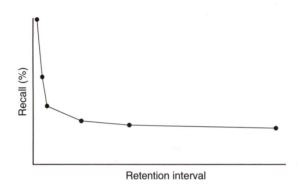

Fig. 1.   A typical forgetting curve

Two main mechanisms of forgetting, from STM or LTM, have been proposed.
Decay theory proposes that memory fades with time. This fading is a basic
biological function, hence altering metabolic rate should alter decay rate and
memory retention. This hypothesis can be tested by altering the environmental
temperature of non-homeostatic (i.e. 'cold-blooded') organisms. For example,
goldfish tend to remember more if their metabolic rate is slowed by cooling the
water that they swim in. Minami and Dallenbach (1946) varied the metabolic
rate of cockroaches by comparing memory in light and dark conditions.
Cockroaches prefer being in the dark and remain inactive; in bright conditions
they are much more active, searching for a dark hiding place. They remembered more when placed in dark conditions than in light.

These findings support decay theory, by showing that forgetting is faster
when metabolic rates (and hence hypothetical decay rates) are faster. However,
they are also consistent with another theory of forgetting – **interference** theory.
Interference theory proposes that forgetting occurs because memories disrupt
with each other. More active animals are more likely to encounter new stimuli
that overwrite existing memories. This phenomenon is called **retroactive** (backward acting) **interference**. Retroactive interference occurs when new learning
disrupts older memories. For example, if you learn some French phrases and
then go on holiday to Spain, learning some Spanish phrases may make it hard
to recall the French phrases you learned earlier. **Proactive** (forward acting)
interference occurs when information already learned prevents learning of new
information. For example, if you change your telephone number, very good
memory for your old number may make it hard for you to remember the new
number. *Figure 2* shows protocols for experiments to demonstrate the two types
of interference.

(a) Retroactive interference

Group	1st study list	2nd study list	Recall	Result
Control	A	nothing	A	good
Experimental	A	B	A	poor

(b) Proactive interference

Group	1st study list	2nd study list	RI	Recall	Result
Control	nothing	B	–	B	good
Experimental	A	B	–	B	poor

Fig. 2. Demonstrating interference in the laboratory. RI = retention interval.

People have claimed that you remember things better if you learn them just before going to sleep. There is some support for this claim, although the beneficial effect of sleep is relatively small. Passive decay at a constant rate cannot explain this effect of sleep, because decay should continue to occur despite sleep. Sleep may improve memory by minimizing retroactive interference, by giving the new information time to be encoded and stored before further information is encountered. The finding that memory is better when learning is followed by slow-wave or deep sleep, rather than rapid eye movement or dream sleep, supports the interference account (e.g. Ekstrand *et al.*, 1977; but see Topic K2 for evidence that dreaming may aid memory consolidation).

If all old and new memories interfered with each other, it would be very hard to remember anything at all. In fact, the amount of interference depends on the similarity of the things you are trying to learn. Very similar items interfere more with each other than more distinct items. Gunter, Berry and Clifford (1981) demonstrated this by asking people to listen to four successive items on a mock news bulletin. For the control group, each item was about the same topic; recall of the later items was worse than for the first item. For the experimental group, the first three items were on the same topic but the fourth item was on a new topic; this group recalled the fourth item of their list much better than the control group recalled their fourth item. This beneficial effect on memory of changing the topic is called **release from proactive interference**. You can demonstrate it for yourself by asking a friend to remember four item lists of fruits until they fail to get a list correct, followed by a four item list of animals (which they should recall with ease).

Encoding

Encoding is the first of three stages of remembering, the other two being storage and retrieval. These three stages interact, hence encoding something well will improve its storage in memory and make it easier to retrieve later. Good encoding involves understanding, organizing and integrating incoming information.

Organizing information by **chunking** can improve retention in STM (e.g. Miller, 1956). Chunking can involve dividing a list into portions separated by pauses, for example, 139 856 247 (think about how you say your telephone number) or dividing it into meaningful units, for example, BMW, HIV, USA. Practicing imposing meaning and structure on material can have dramatic benefits for recall. Ericcson, Chase and Faloon (1980) tested a volunteer who increased his digit span to almost 80 after 230 hours of practice at chunking digits as running times (see also Topic D2).

Although Atkinson and Shiffrin argued that information is transferred from STM to LTM via rehearsal (see Topic D2), Craik and Watkins (1973) demonstrated that mere repetition or **maintenance rehearsal** has little benefit for LTM. They asked participants to listen to word lists and remember the last word beginning with a particular letter. They manipulated the length of time for which a word must be maintained before being superseded by another word starting with the same letter. For example, if participants are asked to remember the last word starting with 's' in the following list, they should rehearse 'star' for longer than 'sugar': blackboard, olive, sugar, star, pit, rose, chair, salad, jacket. On a surprise recall test, the most rehearsed words were remembered no better than the least rehearsed.

In contrast, **elaborative rehearsal**, which involves imposing meaning or structure on the material to be learned, does improve LTM. Thus the locations of stars in the sky are more easily remembered if they are encoded as constellations rather than individually. Many **mnemonic techniques** use imagery to aid memory by integrating and organizing material. For example, orators in ancient Greece used the **method of loci** to remember their speeches. They visualized a familiar route and imagined placing each key point of the speech at a different location on the route. To recall the key points, they mentally retraced the route, picking up the points as they went. Try using this technique to remember your next shopping list. The use of imagery helps us remember verbal material by encoding it in two ways, verbally and visually (see Topic E1). The use of a well-learned route helps to structure the material and provides **retrieval cues.**

Summarizing previous research on memory encoding, Craik and Lockhart (1972) proposed a **levels of processing** framework for understanding memory. They argued that retention of information in memory depends on how deeply that information is processed at encoding. Deep processing involves thinking about the meaning of information (e.g. deciding whether the word 'whale' fits the sentence 'I ate my _____') whereas shallow processing involves thinking about its physical characteristics (e.g. is the word 'whale' printed in capital letters?). Deep (or semantic) processing leads to better memory than shallow processing. A consequence of Craik and Lockhart's approach is that it removes the need to postulate separate short-term and LTM stores. Rather, they argue, apparent STM and LTM effects can be explained in terms of levels of processing, STM tasks typically requiring rather shallow processing, for example, simple maintenance rehearsal, and LTM tasks encouraging deeper, more elaborative, processing of the material to be learned. The concept of levels or depth of processing remains influential. However, as a theory of memory, Craik and Lockhart's approach was less successful because of problems defining what constituted deep (or shallow) processing and avoiding the circular argument that deep processing was the sort of processing that led to good memory. Another problem is that there are exceptions to the finding that semantic processing leads to better memory than shallower processing. If a memory test simply requires free recall of items, then semantic processing at encoding is very effective for later success on the memory test. However, if the memory test provides a cue such as 'recall any words that rhymed with dog', then recall may be better if the original learning task involved thinking about the sounds of the words than about their meanings. Morris, Bransford and Franks (1977) explained such findings by the concept of **transfer appropriate processing**, arguing that the important thing for memory is the match between processing at study and retrieval.

Storage

How can memory be so efficient, that is, how can we retrieve information from memory so quickly and accurately? Answers to this question often focus on the way memory is organized, and in particular how word meanings are stored. For example, Collins and Quillian (1969) suggested that similar concepts are stored closer together than unrelated concepts. Concepts are stored as **local representations**, that is, each aspect of a concept is represented as a single node (see Topic E4). The nodes of related aspects and concepts are linked together in a hierarchical fashion. The links represent relationships between nodes, for example 'is a' or 'has'. The hierarchical arrangement of concepts embodies the principal of **cognitive economy**, of minimizing the number of representations of any one piece of information. *Figure 3* shows an example: we know that canaries have wings because the concept of canary is stored under the higher level concept of bird, the property of having wings being stored just once at the bird level rather than many times at each species level. Quillian implemented this hierarchical model of semantic memory in a computer program that could understand text, the **Teachable Language Comprehender** (TLC). The model correctly predicted that people would be faster to verify statements such as 'a canary is yellow' than sentences like 'a canary has wings', which involved moving up the hierarchy. However, it could not explain the converse pattern of findings for falsification rates: people falsified statements such as 'a canary is a tulip' faster than 'a canary is a robin'. It also failed to explain **typicality effects**, the finding that people are faster to verify category inclusion for typical category members. For example, they verify 'a robin is a bird' faster than 'an ostrich is a bird' (see Topic E3 for more on concepts).

Feature theory (Smith, Shoben and Rips, 1974) explains these data by assuming that concepts are stored as clusters of defining and characteristic features. Statement verification takes place in two stages. Stage 1 is a quick comparison of both types of features. 'A robin is a bird' is verified rapidly because robins share many features with prototypical birds. 'A robin is a car' is falsified rapidly because there is no feature overlap. However, stage 1 processing would

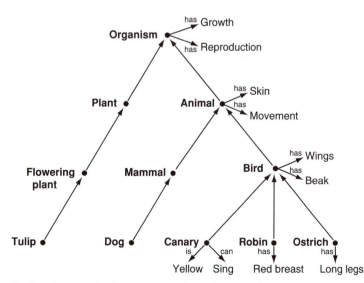

Fig. 3. An example of concept organization in semantic network models of memory such as that proposed by Collins and Quillian (1969)

not give a clear decision in cases where there was just a modest overlap, e.g. deciding whether 'an ostrich is a bird' or 'a bat is a bird'. In this case, the decision is slow because it passes to stage 2, a thorough comparison of defining features.

Although feature theory aims to explain people's **performance**, it says little about their **competence**, that is, about the underlying structure of their cognitive systems. A revised network model, the **spreading activation model** (Collins and Loftus, 1975) aimed to explain sentence verification performance and say something about how semantic memory is structured. The model is similar to TLC in that it comprises a hierarchical array of nodes and links. However, the links have variable associative strengths. The links strengthen with repeated use, so that typical features and concepts are more strongly linked than less typical or less common items. When a node is activated, the activation spreads along the links to related nodes, traveling faster along the stronger links. **Semantic priming** experiments support this concept of spreading activation. Meyer and Schvaneveldt (1971) asked participants to decide whether both words in a pair were real words or non-words. They responded 'yes' faster when the words were related (e.g. bread – butter) than when they were unrelated (bread – paper). Activation of the node for the first word is assumed to spread to the node for the second word if the words are related (i.e. if there are links between them).

Many of these findings can be modeled by **neural networks** with no local structure. Instead of representing concepts at particular nodes at particular locations in 'psychological space', these models represent concepts as *patterns* of activation across a network of nodes. This approach is called **parallel distributed processing (PDP)**. Frequent activation of nodes in tandem strengthens the connections between them, creating stable patterns of activation as concepts are learned (see Topic E4).

Scripts and **schemas** represent a higher level of organization for frequently encountered situations. 'Scripts' (Schank and Abelson, 1977) refers to structured knowledge about specific situations, for example, eating in a restaurant. When a friend tells us about an evening out in a restaurant, they do not need to tell us that they went there because they were hungry, that restaurants serve food, that there are menus and waiters, or that they had to pay for the food. Even if they do tell us this, we do not need to remember it because we share a script for restaurants. Our script for restaurants minimizes the amount of new information we have to remember – we just need to encode the specific details of our friend's account, and not the generic information. 'Schemas' are structured knowledge of more general concepts, for example buying or liking. Scripts and schemas help explain the benefits of knowledge for memory – the more we already know, the less new information we have to encode and, because it fits into our existing semantic memory structure, the better we can retrieve it.

Retrieval

Tests of explicit memory (see Topic D1) require retrieval of information stored in memory. **Recognition tests** (where an item is presented and participants are asked if they remember it from the study phase) involve a judgment of familiarity followed by retrieval of the specific learning episode. If the retrieval phase fails, people may judge that they **know** the item was in the study phase, based on its familiarity, rather than judging that they **remember** it (Gardiner and Java, 1993). Remembering is a controlled process that maps onto episodic memory whereas knowing is a more automatic process (see Topic C4) mapping onto semantic memory. The idea that remember and know judgments reflect different memory processes is supported by their differential sensitivity to variables such as attentional load (which affects remember judgments) and subliminal presentations of related words at test (which affect know judgments).

Recall tests (where participants are asked to list items from the study phase without receiving those items as cues) involve a retrieval phase requiring search and reconstruction processes, followed by a recognition decision about the retrieved information. This **generate–recognize** or **two-process theory** of recall explains why recall is typically harder than recognition (it involves an additional fallible stage). However, it does not explain the finding that people can sometimes recall items that they have failed to recognize. For example, people may fail to recognize Thomas as the surname of a famous writer yet successfully recall the name Thomas when asked to name a Welsh poet whose first name was Dylan.

Recall is easier if there are cues to guide the search process. The **encoding specificity principle** (Tulving, 1972) states that the closer the match between the study phase and the recall phase, the better the cues and the greater the probability of successful retrieval. The cues can be internal, for example depression or alcohol, or external, for example features of the room in which the study and test phases occurred. The facilitation of recall by such cues reflects **mood-, state- or context-dependent learning** respectively. The benefits for recall of reinstating the original context are dramatically illustrated by Godden and Baddeley's (1975) study of scuba divers learning word lists on land or underwater. Despite the stresses of performing the memory test underwater, words originally learned underwater were recalled better underwater than on dry land.

Bartlett showed that people's recollections are partly reconstructed from existing knowledge, rather than being veridically retrieved (see Topic D1). The involvement of reconstruction in recall means that people are suggestible, that their recall can be distorted by leading questions (see Topic D4).

Implicit memory

Recall and recognition tests are referred to as **direct** tests of memory because they directly ask the participant to use memory to complete the test. Memory can also be tested by **indirect** methods, for example, by asking people to complete word stems (study TRACTOR, test TRA-), to guess the identity of fragmented words (-R--TO-) or pictures, or to rate how much they like the test items (this picks up the **mere exposure effect**, the tendency for people to prefer previously encountered stimuli, see Topics K3, L1). These tests are thought to measure implicit memory (see Topic D1) because they show a change in response behavior to previously presented stimuli in the absence of conscious recollection of those stimuli. Implicit and explicit memory are sensitive to different experimental manipulations, leading to the suggestion that they reflect different underlying memory systems. Manipulations of awareness – for example, dividing attention at study or presenting stimuli subliminally – impair explicit memory more than implicit memory. Conversely, manipulations of superficial aspects of the stimuli – for example, changing the font in which words are written or changing from visual to auditory presentation – typically impair implicit memory more than explicit.

D4 LIMITS AND FAILURES OF MEMORY

Keynotes

Good memory	People typically have excellent recognition but poor recall for pictures and faces. Experts have better memory than novices. Surprising, distinctive and emotional stimuli are remembered well.
Autobiographical memory	Although it is generally easier to recall more recent personal experiences, memories from adolescence and early adulthood tend to be particularly well recalled. Infantile amnesia is the inability to recall experiences from before about 3 years of age.
Prospective memory	Memory for intentions (e.g. to attend a dental appointment) is called prospective memory. Intentions can seem difficult to remember because we have to generate our own cues (e.g. write an appointment in a diary) and remember to notice them (e.g. remember to look in the diary or check the time).
Memory across the lifespan	Although young babies can remember things, memory improves as children get older. Memory declines in later adulthood. There is debate about whether these lifespan changes are due to general factors, such as information processing speed, or specific factors such as use of rehearsal strategies. Memory function is impaired in the early stages of Alzheimer's disease.
Eye-witness testimony	The fact that recall involves reconstruction means that it is susceptible to leading questions. Hypnosis may increase willingness to reconstruct memories, increasing recall at expense of accuracy. Eye-witness testimony can be improved by careful questioning using the **cognitive interview**.
False memories	Misleading information can produce 'recollections' of events that never happened. False memories can be created in the laboratory by asking people to remember lists of words that are all associates of another, omitted word. People often recall the omitted word and can be very confident that it was present in the original list. These findings have implications for the debate on recovered memories.
Memory damage	Amnesia and agnosia support the hypothesis that episodic and semantic memory are separate systems. Amnesia is a loss of episodic memory with preserved short-term memory (STM), semantic memory and implicit memory. Agnosia is a loss of semantic memory with preserved episodic memory. Sometimes agnosia is selective for particular categories of information, for example, faces (prosopagnosia) or living things.
Related topic	Forgetting and remembering (D3)

Understanding the ways in which memory normally succeeds and fails is important for understanding the limits of memory in everyday life, for example in eye-witness testimony. Conversely, studies of memory failures following brain damage have helped the development of theories about normal memory function.

Good memory

People typically have excellent memory for visual stimuli – think about how many people you are able to recognize in your daily life. Standing, Conezio and Haber (1970) presented 2560 colored slides for just 10 seconds each. Several days later, participants were able to pick out 90% of the pictures on a **forced-choice recognition test** where each previously presented slide was paired with a new slide. In contrast to recognition, recall of visual stimuli is sometimes very poor. The difficulty of recalling details of someone's appearance, for example the color of their eyes or the style of their jacket, after a brief encounter is a problem for eye-witness testimony.

Expertise aids memory. Appropriate scripts and schemas add meaning to stimuli, and provide organized storage structures and good retrieval cues. An example is memory for the locations of chess pieces on a chess board. Expert chess players can recall many more locations than novices, but only if the locations are from possible games rather than random locations on the board (see Topic F3).

Surprise, shock and novelty can aid memory. Just as an unusual feature 'pops out' of an otherwise uniform visual array, so unexpected items can stand out in memory from a list of more ordinary study items. This is called the **von Restorff effect**. Demonstrate it for yourself by asking a friend to remember a long list of neutral words, in the middle of which is an expletive. The expletive should be recalled better than the other mid-list words (the first and last items of the list should also be well remembered – see Topic D1).

Autobiographical memory

Our ability to recall personal or autobiographical memories is different for different portions of our lifespan. Generally, the more recently something happened, the better we recall it, hence autobiographical memories show a fairly normal forgetting function. However, Rubin, Wetzler and Nebes (1986) showed that 70-year-olds recalled a particularly large number of memories from when they were 15 to 25 years old. This **reminiscence bump** is not peculiar to personal memories – it also occurs if people are asked to recall public events. One explanation is that adolescence is a time of change and of experiencing (or noticing, in the case of public events) many events for the first time. Hence some of the good recall of adolescent memories may reflect a **primacy effect** (see Topic D1) for that type of experience.

Autobiographical memory is influenced by emotion. Emotional experiences are remembered better than more neutral events, and recollection of memories is affected by current mood state, an example of state-dependent learning (e.g. people with depression tend to recall negative experiences; see Topics D3 and L2). An extreme example of the influence of emotion on learning is the phenomenon of **flashbulb memories**. Dramatic, salient, surprising and emotional public events produce apparently very vivid, detailed and long-lasting memories. Brown and Kulik (1977), studying memory in the USA for hearing about the assassination of President J. F. Kennedy, coined the term 'flashbulb memories' and claimed that they were a special type of memory. Conway and colleagues (1994), studying memory in the UK for Margaret

Thatcher's resignation, agreed with this claim. Counter-arguments center on the difficulty of verifying the accuracy of such memories, and on using principles of normal memory function to explain the findings. For example, surprising events are typically well remembered because they are distinctive (the von Restorff effect). Emotional events tend to be particularly well recalled. Dramatic and salient events are probably thought about at length, that is, they undergo deep processing. Publicly salient events are usually discussed frequently, hence the memories are also well rehearsed.

People's earliest recollections tend to be of very emotional or salient events, such as the birth of a sibling. People rarely recall experiences from when they were less than about 3 years old. Several explanations of this **infantile amnesia** have been proposed. One possibility is that the infant brain is not sufficiently developed to encode and store permanent memories before about 3 years of age. Another is that the world of the very young child is so different from the adult world that, although permanent memories are stored, they cannot be retrieved by the adult – an example of context-dependent learning. A third explanation is that young children's lack of language skills or of general knowledge structures prevents them encoding memories in an organized and accessible way.

Prospective memory

People commonly claim to have a 'terrible memory' and back up their claim with examples of forgetting someone's birthday or missing a dental appointment. These are failures of prospective memory, of remembering to do things rather than remembering past events or facts (retrospective memory, see Topic D1). There seems to be little relationship between prospective memory and others types of memory. Kvavilashvili (1987) found that participants who were good at a prospective memory task (reminding the experimenter to pass on a message) were no better on a retrospective memory task (remembering the content of the message) than those with less good prospective memory.

Prospective memory seems difficult because we often have to generate our own cues to remind us to do things, for example writing an appointment in a diary and then remembering to check the diary. Also, failures of prospective memory are often public, so you have to explain to your boss why you forgot to organize a meeting or apologize to your friend for keeping her waiting.

How do we know when to check our diaries or watches to find out if it is nearly time to do something? Ceci and Bronfenbrenner (1985) tackled this question by recording how often children looked at a clock when baking cakes for 30 minutes. Early in the 30-minute period, the children checked the clock frequently. Then there was a period of infrequent time checks, followed by frequent checking again until it was time to take the cakes out of the oven. Ceci and Bronfenbrenner suggested that the initial phase of frequent time checks enabled the children to calibrate their subjective experience of time passing. Frequent checking later in the waiting period ensured they took the cakes out at just the right time.

Memory across the lifespan

Even very young babies can remember things. Rovee-Collier (e.g. Rovee-Collier and Gerhardstein, 1997) demonstrated this by tying a string from a mobile to the leg of a 3-month-old infant, so that the mobile moved whenever the infant kicked. Infants in these experiments soon learned that kicking caused the mobile to move. Up to 2 months later, the infants showed they remembered by kicking as soon as they saw the mobile.

Children's memory abilities improve as they get older, for a variety of possible reasons: (i) increased memory capacity; (ii) increased speed of processing; (iii) improvements in strategies for remembering, for example better use of rehearsal; (iv) improvements in semantic memory, providing a framework for encoding new information (see Topics D3 and J1).

Older adults, particularly those over about 65 years old, tend to perform less well on laboratory tests of working memory, episodic memory and prospective memory than younger adults. The causes of this decline in memory performance are not well understood. It may be due to a general slowing of information processing (Salthouse, 1991), or to deficits in specific aspects of memory such as the ability to rehearse or to encode information deeply.

Tasks like implicit memory tests, that benefit less from deliberate learning or remembering strategies, show less effect of childhood development or adult aging. Thus children and elderly adults tend to be as good as young adults on tests of implicit memory (see Topic D3). Elderly adults tend to perform relatively well on 'real life' rather than laboratory tasks where they can use their knowledge and experience to support performance.

Alzheimer's disease is a form of dementia that particularly affects elderly adults. It is a progressive degenerative disorder that causes memory problems in the early stages and widespread intellectual impairment and personality change in the later stages. Baddeley and colleagues (1986) have shown the people with Alzheimer's disease have particular difficulty doing two things at once (a digit span task and a visuo–spatial tracking task in their study), suggesting problems with central executive function (see Topic D2).

Eye-witness testimony

Although crimes may have the characteristics of surprise and emotion that usually make events easy to remember, it is often very difficult to recall the details such as facial features and clothing that are needed to apprehend the criminal. The fact that recall involves reconstruction means that eye-witnesses are susceptible to **leading questions** about such details. Loftus and Palmer (1974) showed participants a film of a car accident. After watching the film, some participants were asked how fast the cars were going when they hit each other while other participants were asked how fast the cars were going when they smashed into each other. Using the phrase 'smashed into' led to higher estimates of speed and an increase in incorrect reports of seeing broken glass. Loftus argues that misleading post-event information permanently distorts the original memory. An alternative explanation is that, when people cannot recall a piece of information, they respond to the **demand characteristics** of the task and try to please the experimenter by reporting the information just suggested to them, rather than saying they do not know. However, even when participants are warned that the post-event information was misleading, their recall is still distorted by it.

Hypnosis appears to improve memory recall and is sometimes used by police and therapists for this purpose. However, it seems to work by increasing people's willingness to reconstruct missing details using their generic knowledge, hence it increases the amount of inaccurate information recalled (see Topic K2).

Eye-witness testimony can be improved by careful questioning using the **cognitive interview**. Four features of the cognitive interview aim to increase the amount of detail correctly recalled without increasing incorrect responses. Eye witnesses are asked:

- to recreate mentally the context in which the crime took place and how they felt at the time;
- to report everything they can think of relating to the crime, even if it seems trivial;
- to report the details in different orders, e.g. backwards as well as forwards;
- to report the event from the perspective of other people involved, as well as from their perspective.

Asking witnesses to reinstate the original context is a way of tapping context-dependent learning (see Topic D3) and improving accurate reconstruction of the event.

False memories   Leading questions and misleading information not only distort eye-witness testimony but can also help to create completely new autobiographical memories. These new memories are known as **false** or **implanted memories**.

Loftus (1993) reported the case of a 14-year-old boy, Chris, whose older brother told him that when he was 5 years old he had been lost in a shopping mall. Chris believed this fictitious account and, 2 weeks later, his 'memory' for being lost in the mall included details that were not present in the original account, for example that the man who rescued him was gray-haired and wearing glasses. He expressed disbelief when told that his memory was false. Similar false memories can be created using doctored photographs. Wade and colleagues (2002) showed adults fake photographs of themselves on a hot air balloon. After a series of recall exercises, 10 of 20 participants had false memories of the event, which in fact had never happened.

False memories can be demonstrated in the laboratory by asking participants to remember a list of words that are all associates of a keyword that is not itself present in the list. For example, participants might be asked to remember a list containing 'thread, haystack, eye, sew…'. Later, many will recall or recognize 'needle' even though needle was not present in the original list. They will often be as confident of their memory for 'needle' as they are of their memory for other words in the list, and will falsely remember 'needle' even if warned about the nature of the lists. This technique is known as the **Deese–Roediger–McDermott paradigm** after its recent authors (Roediger and McDermott, 1995), who modeled their experiments on a study published in 1959 by Deese.

Studies such as these have informed the heated debate on **recovered memories**. Some people have claimed that, through therapy, they have been able to recover previously repressed memories of childhood abuse and other traumatic experiences. The problem is that, without independent witnesses to these events, it is difficult to determine whether the recollection of abuse is a genuine memory or a false memory created inadvertently by the process, during therapy, of repeatedly recollecting and explaining past events. The studies described above show that we cannot assume that a memory is genuine simply because it is reported in detail and with confidence.

Memory damage   Damage to hippocampal and limbic circuits, through infection, stroke, chronic alcoholism or head injury, produces a memory impairment known as organic **amnesia**. Amnesics show a characteristic loss of episodic memory (see Topic D1), specifically an inability to form new memories (**anterograde amnesia**) accompanied by varying degrees of loss of episodic memories from before the damage (**retrograde amnesia**). For example, H.M. (Milner, 1966) became amnesic after surgery to treat epilepsy. He has been unable to learn the address of the house he moved to after the surgery and cannot remember even very salient information such as news of the death of his uncle. Semantic memory and short-term memory (STM) are spared. Procedural and implicit memory are also spared, hence amnesics can learn new skills, for example, programming electronic memory aids.

Agnosia is an inability to recognize objects (see Topic B5). Cases such as J.B.R. (Warrington and Shallice, 1984) illustrate that it is not a deficit of vision but rather a problem understanding the meaning of stimuli, i.e. a problem of semantic memory. When asked to identify objects from pictures, J.B.R. could identify 90% of inanimate objects but only 6% of animals and plants. Similarly, when asked to define words, he gave accurate definitions for names of inanimate objects but was severely impaired at defining living things (e.g. he said he didn't know what a parrot was). J.B.R.'s problems show that agnosia can be confined to specific classes of information. Another example is **prosopagnosia**, failure to recognize faces (see Topic B6). People with agnosia can often describe the detailed appearance of objects yet fail to perceive their overall identity or function. Thus prosopagnosics can describe someone's eye color, hairstyle and facial expression yet fail to recognize them as their spouse or parent. Agnosics often experience problems in just one modality, hence they may be able to identify objects by touch or recognize someone by their voice.

This **double dissociation** between damage to episodic memory with preserved semantic memory, and damage to semantic memory with preserved episodic memory, supports Tulving's claim of a distinction between episodic and semantic memory. The preservation of STM and implicit memory in these cases suggests that these types of memory are also functions of distinct memory systems. Cases of agnosia for particular classes of object (animals, faces) support proposals that semantic memory is highly structured, with similar concepts being stored closer together than unrelated concepts (see Topics D3 and E3).

E1 REPRESENTATIONS

Keynotes

Concept of mental representation

A representation is a label or symbol that stands for something in its absence. Internal or mental representations are the product of perceptual processes and the means by which we store information in memory. There is debate about whether mental representations are analogical (sharing features with the thing they represent), propositional (in the form of abstract, meaning-based symbols for objects in the world), or distributed as patterns of activation across neural networks.

Propositions

Propositions are hypothetical, abstract representations that symbolize, rather than mentally reproduce, the things they represent. A proposition explicitly represents concepts and the relationships between them.

Mental images

Images are analogical representations, sharing some spatial and sensory features with the things they represent. When people are asked to imagine performing a task, their mental performance often resembles actual performance of the task. For example, they take longer to mentally travel longer distances. Kosslyn argues that images are an important form of mental representation, but critics such as Pylyshyn respond that images may just be by-products of cognitive processes operating on propositions, and that participants in imagery experiments are mimicking how they would perform the task in real life, rather than being constrained to perform in that way by the nature of their mental representations.

Dual coding theory

Paivio argued that there are both verbal and sensory representations, called logogens and imagens respectively. Some stimuli can be processed using both codes simultaneously, and this dual coding route is beneficial for subsequent memory. Thus concrete nouns can be remembered as mental images and as words, making them easier to recall than abstract nouns that can only be stored as verbal representations.

Mental models

Johnson-Laird argued that people can use an abstract form of representation called a mental model. Mental models are generated from propositions and are analogical in the sense that they represent spatial features, but unlike images they do not contain sensory information.

Interacting cognitive subsystems

Barnard's interacting cognitive subsystems (ICS) model comprises nine subsystems, each of which uses information in its own representational code. The type of representation depends on the type of input and the level at which it is processed. Two high-level subsystems represent and process two forms of meaning: factual, propositional information and interpretations of implicit information.

Related topics

Short-term memory (D2)
Imagery (E2)
Concepts (E3)

Goals and states (F2)
Mental models (G2)
Mood, memory and attention (L2)

Concept of mental representation

A representation is a sign, symbol or token that 're-presents' something in its absence. External representations include paintings, photographs, written descriptions, clocks, maps, scale models, sculptures, diagrams and so on. Some of these representations bear some physical resemblance to the object or event they represent. For example, a painting or scale model of a town may retain the structural features and spatial layout of the buildings and streets in that town. Representations that do this are called **analogue representations**. Other types of representation, particularly linguistic representations, do not resemble the represented object or scene in any way. Rather, they serve as discrete tokens of objects and their interrelations. Thus letters and words are arbitrary symbols that bear no physical relation to the phonemes or items they represent. Arbitrary symbols like this are called **symbolic representations**, because they symbolize rather than mimic the thing they stand for.

Internal or mental representations serve similar purposes to external representations. They provide a means of manipulating, exploring and describing things in their absence. Mental representations are created by the processes of perception and are exchanged between cognitive processes, serving as the currency of thought. There has therefore been much debate about the nature of mental representation. Much of this debate has focused on the issue of whether mental representations are **analogical**, that is whether we think in images, or whether they are **propositional**, that is language-like descriptions of the world. With the advent of connectionism and neural networks, a third type of representation has been proposed: **distributed representations** are **patterns of activation** in networks rather than localized symbols.

The debate about the nature of mental representation has been most vociferous between proponents of propositions (e.g. Pylyshyn) versus those of mental images (e.g. Kosslyn). However, several theorists have argued that there are several different forms of representation, which are used selectively depending on the different task demands and different cognitive processes in operation.

Propositions

Propositions are abstract representations that do not share the physical properties of the things they represent. Whereas a picture of a dog chasing a ball would show the dog behind the ball so a viewer could infer what was being chased and what was doing the chasing, a propositional representation of the scene would explicitly state the relationship between the dog and the ball:

CHASE(DOG,BALL)

This notation comes from a logical system called **predicate calculus** and is used as a way of expressing cognitive processes that hypothetically use propositional representations. The first item in the phrase is called the **predicate** and represents relationships between the other elements in the phrase. The first element in parentheses is the **subject** (the thing that acts) and the second is the **object** (the thing that is acted upon). The subject and object are both **arguments** of the predicate. The complete phrase is called a **proposition**.

Predicate calculus is written in uppercase letters to emphasize the fact that the words (CHASE, DOG, BALL) represent concepts rather than mental words. Propositions are language-like, in that they are arbitrary labels for real-world objects and relationships, but they do not use any natural language. Rather, they constitute a universal mental code called **mentalese** (see Topic I3).

The fact that propositions are abstract is one reason why they appeal to some researchers as the fundamental unit of representation. If we assume that different sets of cognitive processes use representations in different codes then we need a way of communicating between those processes. For example, if 'dog' is represented as a visual image of a dog and as the word 'dog', then in order to name a dog when we see one we need a code for translating the image into the word. Propositions provide a common language for this translation. However, it does not seem very economical to have a system that uses verbal, visual and propositional codes, so some authors (e.g. Pylyshyn, 1973) have argued that propositions are the sole currency of cognition.

Mental images

How many windows are there in your house? When asked this question, most people feel that they arrive at the answer by visualizing their house and counting the windows they can 'see' in their mental image. Similarly, most people feel that they describe people or places by forming an image of them and 'reading off' the features of the image. A minority of people report no experience of visual imagery, yet can still perform these tasks. Mental imagery provides us with a way of experiencing novel situations or re-experiencing past events. We can imagine what it might feel like to ride a horse, and use that imagined experience to decide whether to take riding lessons. Imagining a long cold drink might provide momentary relief on a hot day or a spur to stop working and go to the refrigerator. Images are analogical representations, sharing some of the sensory and spatial properties of the items they represent.

Authors like Kosslyn (e.g. 1994, see Topic E2) have argued that mental images are a fundamental form of representation, stored separately from propositions and generated in a medium that, like our view of the real world, has limited spatial extent and limited resolution. Kosslyn's arguments are based on research showing that, when people manipulate mental representations, they do so in a way that mimics interactions with objects in the real world. For example, when people imagine moving a heavy object a certain distance, it takes them longer to perform the task mentally than if they are asked to imagine moving a lightweight object the same distance. When people imagine traveling a large distance, it takes them longer than if imagining traveling a shorter distance.

Critics of the view that mental representations are analogical argue that:

- Although we experience 'pictures in our head', images are actually by-products or **epiphenomena** of underlying processes that use propositional representations of the same information.
- Tasks are performed in imagination in a similar way to real life because people are responding to the **demand characteristics** or **task demands** of the experiment. In other words, they interpret the task instructions as a request to **pretend** doing something, and behave accordingly.

Counter-arguments to these criticisms include:

- It is unclear why mental performance should be constrained by the spatial and temporal features of the real world if people use propositional representations that do not share those features. For example, if someone is asked to imagine walking from the sink to the table in their kitchen, they will probably do that faster than if asked to imagine walking the longer distance from the kitchen sink to the bedroom. Both tasks can be represented by similar propositions: GO(SINK, TABLE) and GO(SINK, BEDROOM). How could propositional coding theories explain the difference in the time taken to perform the two versions of the task? One answer is that propositional-based processes may operate in a stepwise fashion because of the demand characteristics of the experiment. For example, the task of walking from the kitchen sink to the bedroom may be represented as a series of stages: GO(SINK, TABLE), GO(TABLE, DOOR), GO(DOOR, CORRIDOR), GO(CORRIDOR, STAIRS), GO(STAIRS, BEDROOM DOOR).

- Even experiments that are designed to hide their goal show similar results. Response times increase with increasing scanning distance (in the hypothetical image) even when the experimental instructions do not mention imagery or participants are told to expect shorter distances to take longer to scan (e.g. Finke and Pinker, 1982).

Anderson (1978) argued that it is impossible to determine what sort of representations are used in human cognition because performance on any task that claims to be based on images can be mimicked by a different set of processes operating on propositions.

Dual coding theory

Paivio (e.g. 1986) proposed that mental processes use two types of representation: **imagens** and **logogens**. Imagens are the currency of cognitive processes specializing in the perceptual analysis and memory of non-linguistic, sensory information, whereas logogens (see Topic I1) are the currency of the language processing system. The concept of imagens and logogens maps onto the distinction between analogical and propositional coding. There are **referential links** between the non-verbal and verbal systems in Paivio's model, allowing us to visualize named objects and name objects we see.

Evidence for Paivio's theory comes from studies of memory that show better recall of concrete nouns (e.g. table, horse), that can be stored in two ways, than of abstract nouns (e.g. idea, desire) for which there is only one route to recall (see Topic D3). In one set of experiments, Paivio asked participants either to pronounce or to image concrete nouns. Some words were repeated. In the single coding condition, participants performed the same task each time they saw a particular word, that is, they pronounced it on both occasions or they imaged it on both occasions. In the dual coding condition, participants imaged the word on its first presentation and pronounced it on the second, or vice versa. On a subsequent surprise memory test, the dual coding condition led to better recall of the words than the single coding condition. Also, words presented twice in the dual coding condition were approximately twice as well remembered as words presented only once, suggesting additive effects of pronouncing and imaging. There was less benefit of repeated presentation in the single coding condition.

Mental models

Johnson-Laird (1983) used evidence from studies of human reasoning to argue that we use a form of representation called mental models, as well as images and propositions. Mental models contain both propositional and analogical information. They are analogical in the sense that they share structural properties of the world such as spatial layout but, unlike images, they have no sensory content. Johnson-Laird views mental models as a high-level form of representation that saves cognitive processes having to operate with propositions, but which can be translated into the more basic propositional code. Evidence supporting the use of mental models in reasoning is discussed in Topic G2. The concept of mental models is also used in language research, for example to help understand text comprehension (see Topic I2).

Interacting cognitive subsystems

Barnard (1985) proposed a theoretical framework for modeling cognition called **interacting cognitive subsystems** (ICS, *Fig. 1*), which sees all cognitive activity as the transformation of information between different levels of mental representation. Barnard's theory argues that nine different levels of representation are sufficient to model cognition:

- Three sensory levels for acoustic, visual and body state information.
- Two perceptual levels for higher-level acoustic and visual information such as patterns and features (e.g. words, objects) abstracted from the sensory input.
- Two response levels for spoken or limb-based output to the external world.
- Two high-level 'central engine' representations for propositional and implicational information.

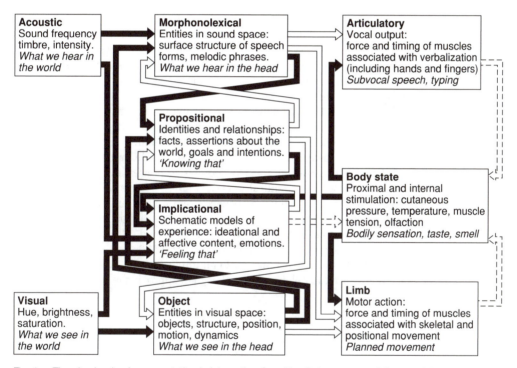

Fig. 1.   The nine levels of representation in Interacting Cognitive Subsystems and the possible transformation processes. Black arrows transform a detailed representation into a more abstract level, white arrows transform an abstract representation into a more detailed level. The dotted arrows operate by making changes to the body so are not cognitive processes.

Each form of representation is processed by a different cognitive subsystem. These subsystems can copy information into their own memory stores and can translate it into different representations, to be processed by another subsystem. For example, the acoustic subsystem receives acoustic representations that contain information about a person's speech, as well as their voice pitch, hisses, coughs etc. It can store this representation, and can translate it into the more abstract, phoneme- and word-based representations processed by the morphonolexical subsystem. If the input is of poor quality, any subsystem can enhance the representations with which it is working by entering **buffered**

processing mode. Buffered processing involves retrieving recent input from the memory store of the subsystem. The retrieved input serves to amplify the information, in the way that replaying an audiotape may help you to understand a poor-quality recording. Buffered processing represents attention focused on the input for that subsystem, so only one subsystem can be in this processing mode at a time, making ICS a blend of specific resource and bottleneck models of attention (see Topic C3).

The two subsystems that form the 'central engine' of the ICS model operate on two different levels of representation, which they continually swap with each other, to maintain an up-to-date understanding of changes in the environment, task demands, goals and so on. The **propositional subsystem** uses an abstract, propositional code to represent explicit meanings. The **implicational subsystem** uses an 'implicational code' to represent information that is more hidden, for example emotional information. Thus one system deals with cold factual knowledge and the other with implicit understanding and feelings. Propositional representations are derived from detailed processing of the abstract content of sensory information, while implicational representations are derived directly from the qualitative nature of sensory representations. The two systems interact so that people's implicit understanding of a task influences their propositional responses, and vice versa.

Scott, Barnard and May (2001) tested the hypothesis that people use both types of representation even when performing an apparently simple task. They asked people to generate a number between one million and 10 million, but they phrased the instruction slightly differently in different conditions. The slight differences in phrasing affected people's responses, even though the set of possible responses was identical in each condition. For example, people who were asked to generate 'a number between one million and 10 million' were more likely to respond with a round million like five million than people asked to generate 'a **random** number between 1 million and 10 million' (who were more likely to respond with a number like 5,000,342). In this condition, the word 'random' in the instructions led to an implication that the number had to be exact, and hence fully elaborated, rather than approximate.

E2 IMAGERY

Keynotes

Imagery and perception	Imagery and perception appear to share some underlying cognitive processes. Imagining a stimulus can help detect that same stimulus in the world whereas imagining a different stimulus impairs detection. Image scanning and mental rotation experiments suggest that people manipulate visual images in the same way that they manipulate objects in the real world. However, mental images differ from external images and objects in the sense that they are interpreted and unambiguous. Kosslyn's theory of imagery proposes that visual images are the form of representation underlying visual perception and object recognition as well as mental imagery tasks.
Imagery and working memory	Baddeley assumed that visuo-spatial short-term memory (STM) and visual imagery were functions of the same working memory system. In general, tasks that disrupt working memory also disrupt imagery in the same modality. However, imagery can be impaired by brain injuries or concurrent tasks that do not affect STM. Logie argued that visual imagery uses the visual storage and spatial rehearsal processes that enable visual STM, but also requires a visual buffer that allows conscious inspection and manipulation of visual representations.

Related topics	Short-term memory (D2)	Representations (E1)

Imagery and perception

Imagery shares some features with perception. When participants are asked to perform tasks on imagined objects, their performance seems subject to the same constraints as performance on real objects. For example, in a **mental rotation** study, Shepard and Metzler (1971) asked participants to decide whether two pictures represented the same object. The time it took them to make the decision was directly related to the degree to which the object in one picture was rotated from the vertical, as defined by the orientation of the object in the other picture. Therefore it appeared that participants were mentally rotating their representation of the stimulus in order to make a decision (*Fig. 1*). **Image scanning** experiments show similar results. For example, Kosslyn *et al.* (1978) taught

Fig. 1. Mental rotation: when asked to decide if a letter is normal or mirror-imaged, people are faster with the left hand stimulus than with the right hand stimulus, which is rotated further from the vertical.

participants the layout of a map of a treasure island. Then they asked them to visualize a feature on the map, say the hut, and asked them to imagine a black dot moving in a straight line from that feature to another feature on the map, say the well or the pond. Participants pressed a button when they had done this. Their reaction times were closely correlated with the physical distances on the map, suggesting that they were performing the task in 'real space' using images that were analogous to percepts of visible stimuli. An alternative interpretation is that performance on these tasks is based solely on propositional knowledge and that participants are responding to the demand characteristics of the task (see Topic E1). These experiments are known as **chronometric studies** of imagery because cognitive processes are inferred from measurements of the time taken to perform the tasks of interest (in Greek, *khronos* = time, *metron* = measure).

Additional evidence that common processes underlie imagery and perception comes from experiments by Segal and Fusella (1970), who showed that auditory imagery (imagining a sound) interfered with detection of faint auditory stimuli whereas visual imagery (imagining a visual scene or stimulus) interfered with detection of faint visual stimuli. Conversely, Farah (1985) demonstrated that asking participants to imagine an 'H' improved detection of a faint letter 'H' (but not a faint 'T'). This **priming effect** supports the argument that imagining and perceiving share common mechanisms.

Brain imaging studies also suggest that similar brain regions are activated during visual perception and imagery tasks. Much of the visual cortex of the brain maps topographically onto the retina. In other words, the part of the brain that receives input from the retina is organized in a way that roughly preserves the structure of the retina. Kosslyn (1994) reported a positron emission tomography (PET) study that suggests visual mental imagery relies on these topographically organized regions of visual cortex. He and his colleagues asked participants to imagine letters of the alphabet, and to make their images as small as possible or as large as possible. When the images were small, the region of visual cortex mapping onto the fovea (see Topic B1) was particularly activated, whereas the larger images were associated with greater activity in visual cortical areas receiving parafoveal input. In other words, the larger images activated areas that would also be activated by looking at large visual stimuli.

However, images do not always behave like percepts. Chambers and Reisberg (1985) briefly showed naïve participants the ambiguous duck–rabbit figure (see *Fig. 1* in Topic A1). They then removed the figure and asked participants if they could reverse their image of the figure to see it as something else. None of the participants could do so, even though they had successfully practiced reversing other ambiguous figures. To rule out the possibility that their images were inaccurate because they had forgotten salient features of the figure, Chambers and Reisberg then asked participants to draw the figure they were imagining. All the participants were able to reverse the figure in their drawing, seeing it as a duck instead of a rabbit or vice versa. This finding suggests that images are meaningful interpretations of stimuli, rather than veridical 'copies' of them.

Kosslyn's (1994) theory of imagery assumes that visual images are the form of representation underlying visual perception and object recognition as well as mental imagery tasks. Kosslyn used data from chronometric studies of imagery and brain imaging studies like those described above to develop a computational theory of imagery in which many different processes act together to give the unified experience of a visual image. Visual images are generated from visual and propositional information stored in long-term memory (LTM). The same information is used to recognize objects in degraded visual scenes; thus, according to Kosslyn's theory, visual perception relies on top-down image-generation processes as well as bottom-up sensory processes (see Topic B5 on change blindness and inattentional blindness). Images are stored in a visual buffer that has spatial properties and a limited spatial extent, hence an image of an elephant is larger than an image of a mouse and, if the imagined elephant is very close to us, the image may be incomplete because it cannot fit within the boundaries of the visual buffer. An attention window focuses on different parts of the visual buffer in the same way you can attend to different parts of a visual scene. Mental rotation and scanning are achieved by transformation processes that act on the information in the visual buffer.

Imagery and working memory

Baddeley (1986) assumed that visual imagery is a function of the visuo–spatial sketchpad (VSSP) of working memory. Support for this assumption comes from a demonstration by Logie, Zucco and Baddeley (1990) that visual imagery interferes more with visual short-term memory (STM) than with verbal STM whereas mental arithmetic interferes more with verbal STM than visual STM. Baddeley and Andrade (2000) showed that working memory is necessary for **vivid imagery**. They asked participants to form visual or auditory images, for example to imagine the appearance of people playing tennis or the sound of people having an argument. During the imagery task, participants performed a concurrent visuo–spatial task (tapping a pattern on a keyboard) or a concurrent verbal task (counting aloud). Then they rated the vividness of their image. The verbal task reduced the vividness of the auditory images more than the visual; the visuo–spatial task reduced the vividness of the visual images more than the auditory. Baddeley and Andrade argued that vivid imagery requires two processes: i) retrieval of sensory information from LTM; and ii) maintenance and manipulation of that information in working memory.

However, neuropsychological evidence contradicts the assumption that visual STM and visual imagery are functions of the same cognitive system. Morton and Morris (1995) reported a patient, M.G., who showed a dissociation between normal spatial STM and impaired performance on imagery tasks such as mental rotation. Laboratory data also reveal differences between imagery and STM. Andrade *et al.* (2002) tested the effect of viewing a flickering pattern of dots on performance of visuo–spatial STM tasks such as remembering the position of black squares in a matrix. Watching the dot pattern had no impact on performance of the STM tasks, even though it is a task that impairs some visual imagery tasks such as using imagery mnemonics (Quinn and McConnell, 1996; see Topic D2).

Thus one debate in the working memory literature is about the extent to which common processes underpin visual STM and visual imagery – are these both functions of the VSSP? Another debate has been about whether the VSSP is essentially visual or essentially spatial. Because it was assumed that the VSSP served imagery as well as STM functions, many of the studies addressing this issue used an imagery task devised by Brooks (1967) and known as the **Brooks' matrix task**. Participants in the imagery condition receive a series of sentences such as 'In the starting square put a 1, in the next square to the *left* put a 2, in the next square *up* put a 3...'. They are asked to remember the sentences by visualizing the path made by the numbers in a matrix. A no-imagery, rote rehearsal condition in which participants receive nonsense sentences like 'In the starting square put a 1, in the next square to the *good* put a 2, in the next square to the *quick* put a 3...' is used for comparison. An influential early study by Baddeley and Lieberman (1980) compared the effects of a spatial task (blindfolded tracking with auditory feedback) and a visual task (brightness judgments) on recall of the spatial sentences in the imagery condition. Recall was only impaired by the spatial task, therefore they concluded that the VSSP uses spatial rather than visual coding. However, subsequent studies have found effects of visual tasks, including brightness judgments, on imagery performance. It therefore appears that visual imagery – and, by extrapolation, visual STM – uses both visual and spatial processes.

Neuropsychological evidence supports the distinction between visual and spatial processes in visual imagery. Farah and colleagues (1988) reported the cases of two patients with contrasting deficits: L.H. performed poorly on visual imagery tasks requiring mental comparisons of size, shape and color, yet had normal scores on spatial imagery tasks such as the Brooks' matrix task; N.C. was impaired on spatial imagery tasks but performed well on visual imagery tasks.

In an updated model of the VSSP, Logie (1995) argues that visual representations are temporarily stored in a **visual cache** and maintained or 'rehearsed' by a spatio–motor **inner scribe**. Thus he distinguishes between visual representations and spatial rehearsal processes. When we need to inspect or manipulate visual representations, they are transferred from the visual cache to a **visual buffer** that is the medium for conscious visual imagery. Thus, in Logie's model, visual STM is a function of the visual cache and inner scribe, whereas visual imagery may use information stored by those systems but also requires a visual buffer for conscious inspection or manipulation of images.

E3 CONCEPTS

Keynotes

What is a concept?	A concept is a unit of knowledge that allows us to recognize novel stimuli as types of thing we have encountered before. Concepts tend to have fuzzy boundaries, so it is hard to say where the category of, say, sport ends and that of games begins. Category members, or exemplars, share characteristic features, giving them a family resemblance to each other. Some exemplars are more typical of the category than others, giving categories a graded structure. This graded structure is unstable: different items are rated as most typical on different occasions, depending on the task demands.
Concept learning	Research into how we learn concepts and store them in semantic memory has used abstract stimuli that give the experimenter control over variables such as distance from a hypothetical prototype. The main theoretical debate has been about whether we store every exemplar or whether we extract a prototype from the training exemplars and just store the prototype.
Prototype theories	Typical exemplars are categorized quickly because they are very similar to a prototype containing all the characteristic features of the category. After training with a set of exemplars, participants categorize the prototype more quickly than novel exemplars. This suggests they have extracted the prototype from the training set. However, they categorize exemplars from the training set fastest of all, suggesting that they still have access to information in memory about specific instances.
Instance theories	Typical exemplars and prototypes are categorized quickly because they share many features with the exemplars or instances stored in memory. Priming studies show that categorizing novel stimuli speeds up subsequent categorization of old exemplars. This shows that old exemplars were retrieved from memory in order to categorize the new stimuli.
Related topics	Forgetting and remembering (D3) Connectionism (E4) Representations (E1)

What is a concept?

A concept is the knowledge that enables us to categorize some things as being of the same kind and other things as being of a different kind. Concepts are important because they enable us to make sense of the world and to know how to deal with novel experiences. Every time we encounter a cat, we receive a different image on our retinas, even if it is our own cat whom we see many times a day. We need a concept of 'our cat' to recognize that each image belongs to the same cat, and a more general concept of 'cats' to identify other cats as being the same type of thing. The concept not only allows us to recognize certain smallish furry animals as cats but also allows us to decide whether to cuddle, eat, or run away from them. The contribution of conceptual knowledge to perception results in **categorical perception** effects, whereby a range of stimuli can be perceived as a smaller number of categories. Stimuli that are perceived as being within a category cannot easily be distinguished from each other, whereas stimuli that are perceived as being in different categories are easy to distinguish, even if the objective difference between them is smaller than that between stimuli within the category (see Topic H4).

Some categories, for example triangles or widows, are relatively easy to define. To be included as members (or **exemplars**) of these categories, things must possess certain **defining features**, for example, a closed shape with three sides and angles adding up to 180 degrees, or a living female human whose spouse has died. Most concepts are less easy to define. For example, most people understand what is meant by the term 'sport' and might agree that tennis and football are sports, yet there is no generally agreed set of attributes that defines a sport. Speed, fitness, teamwork, competition, agility are all **characteristic features** of sports, but many sports have only some of these attributes. The philosopher Wittgenstein used the term **family resemblance** for this tendency of category members to share some characteristic attributes or features while differing from each other on other attributes. In common with members of human families, some members of categories resemble each other more than other members, that is they are more typical of the category. Typicality affects performance on tasks that involve categorization. Examples of **typicality effects** include:

- People are faster at deciding category membership for typical exemplars (e.g. robin as a member of the category 'birds') than atypical exemplars (e.g. emu).
- Children and adults learn about typical exemplars more easily than atypical.
- When asked to generate members of a category, people respond with typical exemplars more often than atypical ones, suggesting that typical exemplars are easier to retrieve from memory.
- People use typical exemplars as reference points in descriptions, for example they might say 'a parallelogram is like a rectangle' but not 'a rectangle is like a parallelogram'.

Rosch (1973) argued that typicality effects are important because they show that concepts have a **graded structure** and fuzzy boundaries, with some members of a conceptual category being more central to the category and more important for cognition than others. Models of how concepts are represented in **semantic memory** need to take this graded structure into account. **Spreading activation networks** do so by assuming that typical exemplars are more strongly linked to their category node than less typical exemplars. **Feature theories** assume that decisions about category membership are made by comparing the defining and characteristic features of the exemplar with those of the concept (see Topic D3).

Spreading activation networks and features theories are examples of **structuralist accounts** of semantic memory. A problem with trying to model the structure of semantic memory is that concepts appear to change their structure depending on the demands of the task in hand. Examples of the **instability of concepts** include:

- People are inconsistent in their use of concepts. For example, someone may decide that snooker and pool are sports when deciding what sort of program to watch on television yet not consider them to be sports when thinking of ways to get fit. A botanist might classify a tomato as a fruit but look for it in the vegetable section of the supermarket.
- Typicality effects are affected by task context. Roth and Shoben (1983) showed that 'cow' was judged as a more typical animal in the context of 'milking an animal' than in the context of 'riding an animal'. The typicality of tea and milk as drinks was reversed depending on whether the context was secretaries taking a midmorning break or a truck driver having breakfast.

Barsalou (e.g. 1982) argued that semantic memory is rather unstructured. Although concepts have a few core properties that are accessed regardless of task context, their apparently graded structure changes according to the goals and constraints of the task in hand.

Concept learning

Research into concept learning aims to discover how concepts are created and suggests ways in which, once learned, concepts may be represented in memory. **Rule induction theories** propose that people are taught or can abstract the rules that govern category membership, and then use these rules to classify new exemplars as category members. Rule induction is unlikely to play a major role in learning fuzzy natural categories, although it is important for learning some artificial categories with clear sets of defining features (e.g. triangles). **Instance theories** propose that concept learning is a matter of encoding each exemplar of a category along with its category label, whereas **prototype theories** propose that concept learning involves extracting a single mental prototype from many individual exemplars.

Studies of concept learning typically use rather artificial stimuli, such as abstract dot patterns. These stimuli have two advantages: i) they are novel for all the participants in an experiment, regardless of their general knowledge of the world; and ii) they allow the experimenters to manipulate variables such as the extent of the difference (or the 'distance') between one exemplar and other exemplars or the prototype. A typical study involves a training phase, in which participants learn to classify many instances of one or more categories, and a test phase in which they are tested on their ability to classify new stimuli as belonging to one or other of the learned categories.

Prototype theories

Prototype theories propose that concepts are represented in memory by prototypes that contain all the salient features of the concept. Typical exemplars of a category are classified rapidly because they are more similar to the prototype. Evidence that people extract prototypes comes from studies showing that:

- In some conditions, participants are more likely to classify the prototype correctly, even though they have never seen it before, than the training instances that they had seen before. This **prototype effect** happened particularly when a large number of instances were presented during training and there was a relatively long interval between training and test (Homa, Sterling and Trepel, 1981).
- On a memory test following a category-learning phase, participants were more likely to say that they recognized the prototype, even though they had not seen it before, than the actual training items (Metcalfe and Fisher, 1986).

This finding suggests that the prototype was better represented in memory than the instances.

However, there are also problems for prototype theory:

- In the Homa *et al.* study, stimuli that were equidistant from the prototypes of two categories were classified on the basis of their similarity to the training instances, suggesting that information about the specific instances was still available in memory. This persistent influence of the training instances is known as the **exemplar effect**. Another example of the exemplar effect is the finding that specific training instances are classified even faster than the prototype.
- People know about correlations between features of category instances. For example, we know that small birds are more likely to sing than large birds. It is hard to see how this sort of information could be encoded in a single prototype for 'bird'.

Instance theories   Instance theories, for example Shin and Nosofsky's (1992) **generalized context model**, propose that we store each exemplar we encounter in memory. Categorization involves retrieving representations of the training instances from memory and computing the similarity of the new item to them. New items are categorized quickly if they are typical of a category because they share features with more of the stored exemplars than atypical items.

Findings that are problematic for prototype theories, for example exemplar effects and knowledge of feature correlations, support the hypothesis that we store specific instances in memory rather than just an abstracted prototype. Further evidence for instance theories comes from **priming studies**. For example, Malt (1989) taught participants two categories and then asked them to classify the original training instances and new instances of those categories. She hypothesized that classifying the new items would involve retrieval of the training instances from memory. Because recently retrieved items remain 'active' for a short time, they should influence performance on the subsequent trial. This is what Malt found. Classifying a new item primed, or sped up, classification of a similar training item on the next trial.

Instance theories can explain prototype effects without hypothesizing that a prototype is stored in memory. A test item that would be similar to the prototype shares many features with the instances stored in memory. Therefore it can be classified quickly and is more likely to be mistakenly recognized as an old item than a test item that is more different from the set of learned instances.

E4 CONNECTIONISM

Keynotes

Distributed representations	Connectionist models exhibit parallel distributed processing (PDP). In contrast to symbolic models, which represent one idea per node, they represent concepts as patterns of activation distributed across many nodes and links. Nodes respond to input in parallel and in turn excite or inhibit the other nodes they connect to.
Learning in connectionist networks	Connectionist models learn by matching the output they generate to the correct output they are shown during training. Mismatch between the two results in an error signal that is fed back into the system, which adjusts the strength of the links between nodes so the match is improved next time around.

Related topics Sub-disciplines of cognitive   Forgetting and remembering (D3)
                    psychology (A4)   Concepts (E3)

Distributed representations

In the type of computational model known as **symbolic models of cognition**, each node symbolizes a different idea. These models are therefore said to use **local representations**, because each idea is represented at a single node or location (*Fig. 1a*). For example, in **semantic network models** such as Collins and

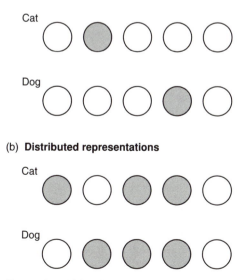

(a) **Local representations**

Cat

Dog

(b) **Distributed representations**

Cat

Dog

Fig. 1.   Local (a) and distributed (b) representations. The five circles represent the same array of nodes in each case. White circles are inactive nodes (i.e., nodes unaffected by the stimulus) and gray circles are nodes activated by the stimulus ('cat' or 'dog').

Quillian's (1969) model (see Topic D3), each node represents a category label (e.g. bird) or a feature or instance of that category (e.g. wings, canary). The nodes are connected by links specifying their relationship (e.g. has, is a). Another type of symbolic model involves linking representations by a series of IF … THEN rules that specify the output required if given a particular input. These models are known as **production systems** (a 'production' is an IF … THEN rule). The productions, which are stored in long-term memory, monitor working memory for the stimulus specified by their IF component. Examples of production systems are SOAR (Newell, 1990) and ACT* (Anderson, 1983).

In contrast to symbolic models, most connectionist models represent ideas as patterns of activation distributed across many nodes and links, that is they use **distributed representations** (*Fig. 1b*). Each node may contribute to the representations of many different concepts. Nodes have multiple connections, some of which excite or pass activation onto connected nodes, and others of which inhibit or reduce the activation of connected nodes. In this sense they are like the neurons in the brain, so connectionist models are often referred to as **neural networks**. The connections between nodes have modifiable strengths or **connection weights**. Nodes receive input and pass on activation in parallel (all at the same time), hence the term **parallel distributed processing** (Rumelhart and McClelland, 1986).

Learning in connectionist networks

In the simplest form of connectionist network, stimuli activate nodes in an **input layer** which in turn activate or inhibit nodes in an **output layer**. The model learns through a training phase in which the response of the output layer, that is the pattern of activation across the nodes, is compared with the desired response provided by the programmer. The difference between the two is called the **error signal**. The connection weights of the links between the input and output nodes are adjusted so that the output becomes more similar to the desired output on subsequent trials. Different models use different formulae or **learning rules** for adjusting the connection weights. One of the most common is called the delta rule. The process of computing the mismatch between actual output and desired output continues over many training trials until the network reaches equilibrium, in other words until it consistently gives the correct output and needs to make no further adjustments to the connection weights.

Two-layer networks are too simple for solving most of the interesting problems in cognition. Greater flexibility and a better match to human cognitive performance can be provided by adding a third layer of nodes with connections to both the input layer and the output layer (*Fig. 2*). This third layer is called the **hidden layer** because it makes no connection with the outside world (it does not respond directly to stimulus input or produce a response). During the training or learning phase, information about the mismatch between the network's response and the desired response is fed back into the network by a method called **back propagation**. The hidden layer functions as a feature detector, extracting common patterns of activation from the input layer and using these patterns to influence the response made by the output layer.

Once it has been trained, a connectionist model should be able to make appropriate responses to novel stimuli that were not part of the training set. For example, if it has been trained to recognize Rover and Fido as dogs, it should also be able to recognize Lassie as a dog even if it has never 'seen' Lassie before. This is the critical difference between connectionist and symbolic

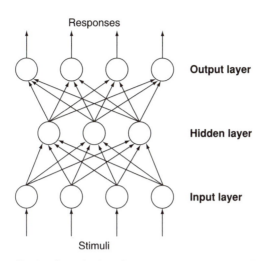

Fig. 2.   Part of a three-layer connectionist network. The circles represent nodes and the arrows between them represent connections.

systems. Connectionist systems can learn information that is not directly programmed into them, and human-like behavior can emerge from the way the system organizes itself during this learning phase. In contrast, symbolic systems must have all their knowledge programmed into them. Connectionist networks also exhibit a property known as **graceful degradation**. If part of the system is damaged, the performance of the system is impaired in proportion to the extent of the damage, rather than catastrophically. Neural networks can therefore provide good simulations of the effects of brain damage in humans. However, because they do not manipulate symbols, connectionist networks do not constitute models of human information processing in the traditional sense (see Topic A1).

F1 GESTALT PROBLEM SOLVING

Keynotes

Gestalt	Gestalt psychologists saw problem solving as the 'closure' of a problem, achieved by the representation of the problem in an appropriate way. In analogy with Gestalt theories of perception, a problem is only a problem because it is incomplete; the solution makes it complete, and finding the solution 'closes' the incompleteness. Closure is accompanied by the flash of insight or 'aha!' experience.
Representation	Problem solution involves mentally forming and reforming different representations of a problem until the right form is chosen. The ease with which the appropriate representation can be found changes the difficulty of the problem. Difficult problems are those in which the appropriate representation is not apparent from the initial description and must be discovered by the solver, or where inappropriate but conventional representations are presented, and must be ignored by the solver.
Set	Set (or fixity) is a tendency to keep thinking about familiar uses of objects within a problem (functional set), or about familiar approaches to solving a problem (operational set), even though they are not helping you find a solution. Set can be reduced by giving objects nonsense names, or by intentionally trying to think of novel uses for objects.
Incubation	After working on a problem unsuccessfully for some time, people put it to one side for a while and engage in a routine or relaxing task. During this phase of incubation, the solution may appear, as if in a flash of insight. Incubation may allow set to weaken, by giving time for obvious (but wrong) ideas to fade; or it may give time for unconscious processes to continue re-representing the problem. It is important for the time to be filled with activity that is mentally non-demanding.
Insight	Insight has been equated with the moment of creative inspiration. Productive thinking is an insightful mode that allows novel associations to be made.
Related topics	The history of cognitive psychology (A3) Visual form perception (B4) Goals and states (F2)

Gestalt

The Gestalt psychologists' general approach to psychology was that mental activity represented the completion by the brain of the fragmentary representation of the world provided by the senses (see Topic B4). In perception, a 'good form' such as a circle would be perceived and recalled even if the original stimulus was not exactly circular, or if it contained a small gap – a process known as **closure**. Problem solving occurred in an analogous way. A problem was only a problem because it was not complete; it needed a solution. Arriving at a solution was equivalent to making the problem complete. Difficult problems were those where the way that the information was represented did not support closure, possibly because of the way that it was represented in the mind. The process of problem solving involved thinking about the problem in different ways, until it was represented in a way that allowed closure to occur.

One particular aspect of problem solving that interested the Gestalt psychologists was the 'flash of insight' or the 'aha!' experience felt when a solution apparently occurred to the solver out of the blue. The solver might have been working on the problem for some time, considering all sorts of possible solutions without getting anywhere, but when the correct solution occurred to them, it did so suddenly, without their being aware of slowly realizing little pieces of it. The flash-of-insight experience is common in anecdotal accounts of great instances of real-life problem solving, as in Archimedes' discovery of the principle of the displacement of an equal volume of water by an immersed object, which lead him to cry 'eureka!' (ancient Greek for 'I have found it!'). The Gestalt account explained insight as being the moment when the mental representation of the problem allowed closure: a previously insoluble problem becomes solvable as soon as it is represented in the right way, and closure occurs instantly. Problem solution involves mentally forming and reforming different representations of a problem until the right form is chosen.

Representation   The Gestalt approach to psychology was primarily based in Germany, and was disrupted by the rise of Nazism and the wars of the 20th century. It also lost popularity as an explanatory account because of the difficulty in finding a neural basis for closure, and because of the increasing emphasis on observable aspects of behavior. Their emphasis on the importance of mental representations became popular again with the rise of cognitive psychology (see Topic E1). Many problems first described by Gestalt psychologists were used again by cognitive psychologists interested in discovering the processes and operations underlying the creation and change of mental representation, for example Maier's (1931) **two string problem** and Duncker's (1945) **candle problem** (*Figs 1* and *2*). In both of these problems, the conventional use of some everyday objects has to be put to one side and an unconventional use for them has to be thought of. The solution to these problems is given at the end of the section, but both require one of the objects to be used in an unusual way. Because most objects have everyday uses, these block the unusual representation of them, a phenomenon known as 'set' (see below).

Kaplan and Simon (1990) investigated the ease with which the crucial aspect of a problem could be made easier or harder to identify by manipulating the representation of the 'mutilated chessboard' problem (*Fig. 3*). A standard 8×8 chessboard has had two opposite corners removed, leaving 62 squares. The problem is to decide whether or not it is possible to lay out 31 dominoes to exactly cover the mutilated chessboard. The answer is that it is not, because the two squares that have been removed are both the same color, leaving 30 white and 32 black squares. Since each domino can only cover two adjacent squares, and adjacent squares are always of opposite colors, the last domino will always have to cover two non-adjacent black squares. In their experiment they used four different versions of the chessboard, intended to make this aspect of the problem easier or harder to discover. The first version used the standard chessboard. The second had the black squares colored white, so that the 'parity' of adjacent squares, and

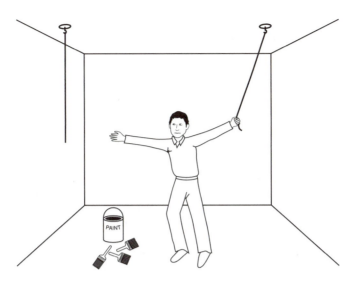

Fig. 1. Maier's Two String Problem (Maier, 1931): the two strings dangling from the ceiling are too far apart for the decorator to grasp them both at the same time; how can he use the objects in the room to help him tie them together?

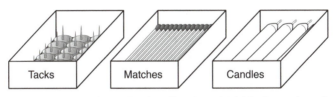

Fig. 2. Duncker's Candle Problem (Duncker, 1945): You have a box of tacks, a box of matches and some candles. Using these items alone, fix a candle securely to the wall so that it can be lit without its wax dripping onto the floor.

Fig. 3. Mutilated chessboard problem (Kaplan & Simon, 1990): This chessboard has had two opposite corners cut off. You have 31 dominoes to cover all of the remaining squares on the board, by placing the dominoes so that each one covers two adjacent squares (horizontally or vertically, not diagonally). Can you either find a solution, or prove that it is impossible?

the 'non-parity' of the removed corners, was not represented at all within the problem description. This made it much harder for people to solve the problem. In a third version, all of the white squares were colored in pink, to draw attention to the fact that color was an important part of the problem and not just something that happened to be there because a chessboard was being used, and in the fourth, the squares were labeled 'bread' and 'butter' to make the parity, and the different number of breads and butters, even more obvious. Both of these were much easier to solve. The more obvious the representation of parity, the sooner it was noticed, and the quicker the solution.

Set

In the descriptions above, one of the explanations for why an appropriate mental representation of a problem is sometimes hard to reach was based on the everyday uses of objects in the problem description 'blocking' the discovery of new uses; another was in the labeling of parts of the problem description which either disguised or drew attention to the crucial aspect of the problem. Both of these relate to the phenomenon identified by the Gestalt psychologists as **Einstellung**, translated as **set** or **fixity**. Set can take two forms: functional set and operational set. Functional set is a fixation upon the common function of objects in the problem, as in the Maier and Duncker problems; operational set is a fixation upon a common way of solving a problem. Luchins (1942) described operational set in his **water jug** problem, which involves being repeatedly given three jugs of different capacities (e.g. 200 cc, 1300 cc, 50 cc) and being asked to use them to measure out an exact amount of water (e.g. 1000 cc). Although the amounts differ on each trial, the problem can always be solved by filling the largest jug first, then using it to fill the middle jug once, and the smallest jug twice, with the required amount being what remains in the largest jug. Although the first couple of trials might take people a minute or two to solve, once they have done so they can easily transfer this solution to the subsequent problems and solve them rapidly. One of the later problems uses quantities such as these, though: with jugs of 23 cc, 49 cc and 3 cc, obtain 20 cc. While the 'formula' still works, there is an easier solution (left for you to discover!). Operational set is displayed when people apply the long solution instead of the short solution. In effect, they were in a 'rut', and applied a learnt method instead of solving the problem afresh. If the learnt method had not worked, operational set might lead them to continue to try to apply it, thinking that they had made a mistake rather than realizing that the problem required fresh insight.

Functional set can be reduced by encouraging people to think about the problem elements in different ways before starting the task. For example, trying to come up with as many different uses for the objects in the problem, or by giving them nonsense names so that their original uses are not made so obvious. The novel uses do not have to be related to the problem, and ideally no attempt should be made to make them relevant. Just thinking of novel uses should be enough to reduce set.

Incubation

Wallas (1926) described problem solving as having four stages. First comes **preparation**, when you discover the problem and think about it unsuccessfully, often for a long time. This is followed by a period of **incubation**, when you give up and do something else for a while, perhaps something relaxing to 'take your mind off the problem', such as Archimedes' approach of going to his local baths. Third comes a brief **illumination** stage, when the flash of insight presents the solution to you, and it is followed by a fourth stage of **verification** as you check that your solution works, and perhaps refine it. In this account, paradoxically, the crucial stage is incubation, in which you are not trying to solve the problem.

Some accounts of why incubation might help have focused on the tendency of good problem solvers to undertake menial or routine tasks (such as doing the ironing or walking the dog) that allow their mind to wander, and become receptive to new ideas – effectively reducing operational and functional set. Another early suggestion (e.g. by Kraepelin, in the 1890s) was that it allowed the 'dissipation of inhibition' or the fading away of obvious but wrong ideas. This apparently fits with more recent cognitive ideas about **spreading activation** (see Topic D3) and connectionist network models (see Topic E4). Yet another possibility is that unconscious elaboration of the problem continues, and that it 'pops into mind' as an insight when the elaborated representation reaches closure.

Silveira (1971, cited by Anderson 2000) studied the effect of incubation by giving people an insight task and then after 15 minutes giving them another task to do for 30 minutes, before resuming the insight task for another 15 minutes. A control group were given 30 minutes to do the insight task without a break. The number of people who could solve the problem rose from the control group's 55% to 64% in the experimental group. Following a break of four hours, 85% solved it in their second 15-minute session. Clearly, the time away from the problem improved their performance. Murray and Denney (1969), however, found that incubation (in the form of five minutes on another task) only helped 'low ability' volunteers, and hindered 'high ability' volunteers. Presumably, the low-ability group were helped to break free from set by the incubation period, while the high-ability group had already done so and were distracted from their incubation by the second task. The best way to make use of incubation would seem to be to concentrate on understanding the details of the problem and checking that you are not taking anything for granted or missing any ambiguities in the description, and to try out all of the obvious approaches for a while. If none of these work, put the problem aside and work on something relaxing and routine, but not something mentally demanding.

Insight

The Gestalt psychologists equated creativity with the illumination stage of problem solving. Wertheimer contrasted '**productive thinking**', an insightful mode which goes beyond the bounds of existing associations or ideas, and '**reproductive thinking**', which is based on what is already known. Smith (1995) suggests that the abruptness of the insight experience is due to a sudden release from set, and that this is more likely to occur after incubation in a context different from the one in which the set was formed (see Topic L3 for the effects of mood on set). Finke (1995) distinguishes:

- **convergent insight**: identifying a pattern or structure within information;
- **divergent insight**: exploring the uses to which a known structure can be put.

Divergent thought is often incorrectly portrayed as inherently more creative than convergent thought, but it is possible to be creative in terms of novelty without producing anything useful: truly creative acts require the novel product to be useful. This is why the Gestalt psychologists emphasized the importance of their final stage, verification. The ability to check that an apparent solution actually does solve the problem, and to try again if it doesn't, is what characterizes truly creative people (see Topic F3).

Note: Solutions to Gestalt insight problems (Topic F1)

Maier's problem: A paint brush (or a tin of paint) can be used as a pendulum weight so that one string can be set swinging back and forth into the reach of someone holding the other string.

Dunker's problem: The solution relies on recognizing that the boxes are also available for re-use, in this case as a shelf to hold the candle and to catch any drips (*Fig. 4*).

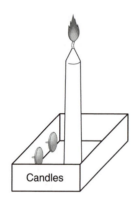

Fig. 4.   Solution to Duncker's Candle Problem (see Fig. 2).

F2 GOALS AND STATES

Keynotes

State space	Newell and Simon (1972) defined state space as including all possible states of all of the entities in the problem. Every state of the problem, including the **start** and the end (the **goal**) are positions in state space. As entities within the problem are changed, the problem's location within state space moves. Problems are solved by finding a solution path that links the start state to the goal state. Problems that have identical state spaces despite different descriptions (surface structure) are isomorphs of each other.
Operators and procedural knowledge	Operators change the states of one or more entities in the problem, so moving the current state in state space. The more familiar we are with operators, the more procedural knowledge we have about them and the easier they are to apply. Problem isomorphs can vary in the amount of procedural knowledge they allow us to use, and so can vary in difficulty despite having the same state space.
Problem types	Problems with well-defined objects, states, operators and constraints are 'well structured', and are amenable to algorithmic solution processes. Insight problems and real-world problems are usually 'ill structured', especially when the exact goal state is not known in advance. These require heuristic solutions, which are faster to apply than algorithmic solutions, but which are not guaranteed to result in the best (or indeed, any) answer. Heuristics include forward and backward searching, generate-and-test and means–end analysis.
Bounded rationality	Simon (1957) suggested that our ability to reason rationally was limited by our knowledge and cognitive capabilities, such as limited capacity memory. This means that while we do attempt to reason in a rational manner, there are bounds placed upon our ability to do so, leading to errors and biases. 'Satisficing' is a problem-solving strategy which leads to an answer that is satisfactory and sufficient rather than optimal.
Related topics	Representations (E1) Expertise (F3)

State space

With the rise of **artificial intelligence** (see Topic A4) came the attempt to program computers to solve problems, both for practical reasons (to build decision support tools, and to avoid the need to have human experts on hand for all decisions) and for psychological reasons (to test that theoretical explanations of problem solving actually worked). Newell and Simon (1972) introduced the concept of a state space, defined as including all possible combinations of the states of all of the entities in the problem. For example, if a problem contains three objects, each of which could be in four different states, then the entire state space has $4 \times 4 \times 4 = 64$ states. The **start state** would be just one of these, defined by the state of each of the objects at the start of the problem, and the **goal state** would be the one in which each object had the required state for a solution. If all three objects had to be in a particular state, then only one of the 64 states would match the goal state; but if (say) one of the objects was not crucial, then the goal space would consist of four alternative states. In this conception, problem solution is a process of finding the shortest (or easiest) path through the state space from the start state into the goal state space (*Fig. 1*).

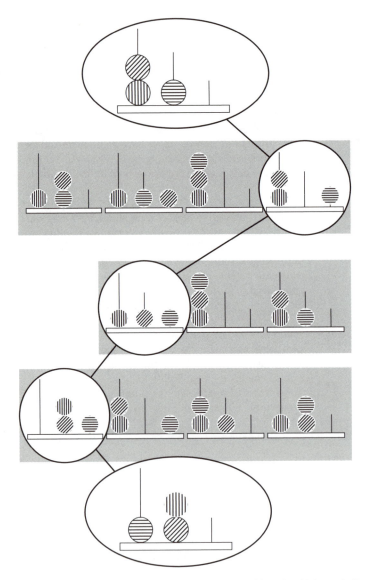

Fig. 1. A search space for a 'Tower of London' problem, in which one ball can be moved on each step of the problem, and the task is to move from the start state at the top to the goal state at the bottom of the figure. On each row, all possible states reachable from the previous state are illustrated, and the correct states highlighted. The path lining them is the solution path.

For example, the **missionaries and cannibals** class of problem involves getting three missionaries and three cannibals across a river in a small boat that can only hold two people at a time. A constraint is that it is not safe to leave more cannibals than missionaries on either bank of the river, or the missionaries will be overpowered and eaten. In this problem, there are really three objects: the number of missionaries on the left bank, with four states ranging from zero to three; the number of cannibals on the left bank, also with four states; and the location of the boat, with two states, on the left or right banks. This means that the problem space (illustrated in *Fig. 2*) has a total of $4 \times 4 \times 2 = 32$ states. If it were not for the cannibalism constraint, the shortest path would involve nine boat trips, since someone always has to row the boat back from the far bank, but there are many different paths because it is not crucial who is on the boat each trip. It is therefore easy to find a solution, because there are no 'dead ends'

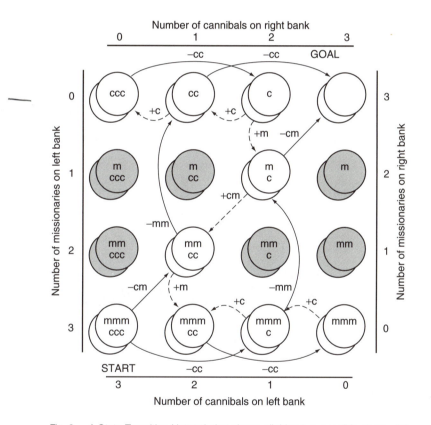

Fig. 2.   A State Transition Network that shows all thirty-two possible states of the Missionaries and Cannibals problem, and the shortest solution paths (taking eleven steps). Each circle shows a different state of the problem in terms of the numbers of missionaries and cannibals on the left bank. Grey circles show states where cannibals outnumber missionaries on one of the banks, and so which must be avoided by a valid solution path. The upper plane of circles has the boat on the left bank, the lower plane has the boat on the right bank. Arrows represent operations (crossing the river in the boat), with solid arrows showing moves from the left to the right bank (upper to lower plane), and dashed arrows moves from the right to the left bank (lower to upper plane). In a valid solution, dashed and solid arrows must alternate. There are two alternative ways of starting (and finishing), and the solution path is symmetrical around the halfway point (two missionaries and two cannibals and the boat on the right hand bank).

in the state space: every move that takes you closer to the goal state is helpful. With the cannibalism constraint, twelve of the possible states must be avoided, and so there are many fewer paths, the shortest solution now taking eleven steps. It is difficult to solve because at one point, you have to move 'away' from the goal state, by making a boat trip that takes two people from the far bank back to the first bank. While this apparently takes your solution path in the wrong direction, it is the only way to work 'around' a state that you are not allowed to enter, where there will be more cannibals than missionaries on one bank.

There are many different versions of this problem, some changing the characters involved and the obstacle to be crossed. One form, studied by Thomas (1974), involves Hobbits and Orcs, traveling through a narrow cave with two lanterns. These different forms of the problem are called **isomorphs**, and can be solved by the same solution path. Because of the similarity in the attributes of the isomorphs, the relationship is easy to identify. Other problems, such as the **Tower of Hanoi** (in which different-sized disks must be moved from one post via a middle post to a third post, with larger disks never being placed on smaller disks, see *Fig. 3*) have isomorphs that are less easy to recognize (e.g. three monsters holding balls of different color, exchanging balls such that the ball in their right hand is always shinier than the ball in their left hand, and if they have a ball in their right hand, they cannot let go of the ball in their left hand). Nevertheless, because the problems are isomorphs, they share structurally identical state spaces, and the solutions are logically equivalent.

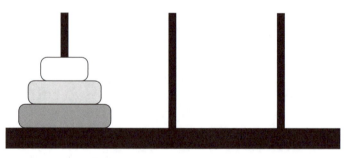

Fig. 3.   The Tower of Hanoi problem. The disks must be moved onto the right hand post, one at a time. No disk may be placed on top of a smaller disk during the task.

Operators and procedural knowledge

The problem is moved through the state space from start to goal states by the application of operators. These are defined as processes which change the state of one or more of the objects in the problem space. In the case of the missionaries and cannibals problem, there are two operators, both taking the form of a boat trip. One operator moves one (or two) people and the boat from the starting bank to the far bank; the other brings one (or two) people and the boat back again. The identities of the people to whom the operator is applied are parameters of the operator. The operators are not always applicable; at any one time, only one of them can be applied, depending upon which bank the boat is on. This makes choosing the path tricky, because applying the operator to get the boat back to the starting bank to pick up two more people always involves moving a person back too, moving you away from the goal state.

In this example, the limited number of operators and the lack of choice over which one to choose at each state simplifies the problem. Increasing the number of operators can make it easier to move from state to state within problem space, but if the constraints get more complicated too, the problem can rapidly get harder and harder. In addition, the operators in this case are easy to apply – you just say 'take the boat' and two of the people do so. In other problems, though, some operators may be harder to apply (find the square root of 169, please!) than others (now multiply 13 by 13) even though they involve the same states. Knowing 'how to do' an operator is called 'procedural knowledge', and if you have more procedural knowledge for some operators than others within a problem space, you may be more likely to construct a solution path that relies upon them rather than the others. This may limit your ability to find the best (or only) solution path.

The way that the problem is represented can also affect the amount of procedural knowledge that you have for a problem, and so affect its difficulty. The standard version of the Tower of Hanoi is easy to understand, because moving disks from post to post is easy to learn, and the constraint about size also conforms to our experience. The isomorph of the problem involving monsters and shining balls does not conform to any procedural knowledge, and so it is much harder for us to apply operators. This makes the whole problem much harder, despite the state spaces being identical, because we have to keep the operators in working memory (see Topic D2) as well as the problem state and the goal state.

Problem types

The missionaries-and-cannibals and Tower of Hanoi problem classes have easily identifiable objects, with precise and discrete states, and the way in which the states of objects change is well specified. The start state and goal state are precisely defined, and when there are constraints, they also allow a known area of the state space to be placed off-limits. Such problems are known as '**well structured**', and they allow an **algorithmic** or step-by-step approach to problem solving, with all possible routes through the space being explored in a systematic way, so that you never get lost, or repeat yourself, and when you have finished, you can be sure that you have found the shortest path. Sequences of operators that are frequently used in order can be 'chunked' into a 'routine' (e.g. in the Tower of Hanoi, to swap the posts two disks are on, move the first disk to the third post, move the second disk to the first disk's old post, and then move the first disk to the second disk's old post); and routines can include themselves, a process called **iteration**.

Other problems, such as the Gestalt insight problems (Topic F1), are not so well defined, and real life problems rarely are. You might not know exactly what the goal state is, nor where the 'boundaries' of the problem are (what objects are part of the problem, and which are irrelevant), nor what all of the applicable operators and constraints are. Such problems are '**ill structured**' and are less well suited to algorithmic approaches. Instead, **heuristic** approaches to problem solving are required. Where algorithms are guaranteed to produce the right answer, heuristics are not, but they are likely to be much faster to use and so more likely to result in a solution, if not the best solution. Heuristics may also be the best approach to problem solving when the problem space is very large, and so when an algorithmic approach would just take too much time. Chess, for example, could be played using an algorithmic approach (simply consider all possible moves you can make, followed by all possible moves your opponent can make; and repeat until all possible chess games have been considered. Now choose the best one!) but because of the number of different states it is computationally impractical. Even computers that have been programmed to play chess use heuristics to reduce the state space that they have to search, by identifying areas as worthless or as against constraints (only moving pawns, for example, or losing too many pieces). Most importantly, they evaluate the positions each state represents as they go, and see how close it is to the goal state (i.e. a win for the computer). Only the search paths that include the best positions are included. The more powerful the chess computer, the more paths it can consider at a time, and the further ahead each path it can look before having to decide which one to choose; but the better a chess computer is, the better is the assessment of the positions at each step, and the less time it spends evaluating paths through irrelevant areas of state space.

Newell and Simon defined four different classes of problem-solving heuristic:

- **Forward search** is the simplest: like the chess example, you start at the start state, and identify the move that makes the most progress towards your goal, then look for the best move from that point, and so on; if you get to a dead end you can mentally undo a move. When you find a path that gets you to the goal, you take it.
- **Backwards search** is similar, except that you start at the goal state and work out what state you would have been in to have got to the goal in one move, and work backwards until you reach your start state; then you retrace the steps.
- **Generate-and-test** is like forward searching, in that you generate alternative courses of action, and see which ones work best, except that you actually take the steps as you go (there is less to remember this way, but it is harder to retrace your steps).
- **Means–ends analysis** compares the current state with the goal state and tries to minimize the differences between the two.

A weakness with the latter two is that they may get stuck in 'local maxima', i.e. they may work towards states where there is minimum difference from the goal state, but which actually then require a large number of steps to actually reach the goal state. A quicker solution might have been possible by avoiding the tempting minimally-different state (*Fig. 4*).

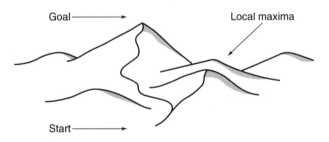

Fig. 4.   The 'local maxima' problem occurs in problem solving strategies such as 'hill climbing', where on each step the move is made that takes you closest to the goal (i.e., takes you highest). At the fork in this path, taking the steepest route takes you to the top of one of the foothills, not to the summit. Each of the foothills is a 'local maximum', but none of them are the 'global maximum'.

| **Bounded** | People have difficulty solving logical problems such as syllogisms (see Topic |
| **rationality** | G1), and in decision-making tasks, where they seem to reason according to |

**Bounded
rationality**

People have difficulty solving logical problems such as syllogisms (see Topic G1), and in decision-making tasks, where they seem to reason according to heuristics rather than precisely logically. This has led some people to conclude that we do not reason rationally at all. Simon (1957) proposed that we do actually try to solve problems and reason rationally, but that our ability to do so is limited by both our imperfect knowledge and our limited cognitive capacities. This means that we are unable to hold all of the relevant information in working memory, or are unable to retrieve relevant information from long-term memory, or we cannot combine all of this information using problem-solving strategies within the time limits imposed upon us by the situation. In effect our rationality is bounded by these limits. When the problem exceeds these bounds, then we have to rely on heuristics and short cuts, and make assumptions that lead to biases and errors. In decision making, Simon (1983) suggested that instead of attempting to maximize our gains and minimize our losses, we use a strategy called **satisficing**, which results in finding the answer that 'satisfies' and 'suffices' even if it is not necessarily the best answer we could find if we worked on the problem for as long as possible. The costs of working on the problem have to be taken into account: it may not be worth working on a problem for another hour just to gain a slightly better answer, when you have already found one that is satisfactory and sufficient.

F3 EXPERTISE

Keynotes

Special knowledge	Experts not only have more knowledge about their area than other people, they have also learnt to structure it into chunks of schematic knowledge. This allows them to encode the surface features of a problem faster than novices, and to build more complex representations. The better the representation, the faster the problem-solving strategies can be applied and the quicker a solution can be found.
Domain specificity	Expertise is usually restricted to a particular area or domain. The schematic knowledge is encoded in terms of features of this domain, and so cannot usually be generalized to other domains. The application of problem-solving strategies that have been learnt also depends upon the problem matching the normal pattern within the domain. Changing the context of the problem means that the expert may not be able to apply their knowledge, even though the problem is essentially the same.
Analogical reasoning	Sometimes experts can apply knowledge to novel situations by making use of analogical reasoning. This involves noting similarities between the novel situation and a previously solved problem, and then mapping the known solution onto the new problem. Sternberg has argued that analogical reasoning is a fundamental aspect of human intelligence. Gentner and Gentner have proposed a structure mapping model that involves the representation of scenarios as a structure of propositional facts; analogies can be discovered when two scenarios can be represented using similar structures.
Expert systems	Computer programs making use of chunking, schematic knowledge and problem-solving strategies have been written for well-structured domains of expertise. They combine facts to make inferences about the situation, and can use this knowledge to search for new facts. They are useful in situations where there is a large amount of information, not all of which is relevant. Finding out what the human expert knows is not straightforward, because their schematic knowledge is not easily verbalizable.
Creativity	Creative individuals are experts who have built up a great deal of knowledge in their field, and who are able to apply ordinary problem-solving skills to their extraordinary knowledge.
Related topics	Forgetting and remembering (D3) Emotion, perception and Limits and failures of memory (D4) performance (L3) Mental models (G2)

Special knowledge

The everyday idea of what an expert is centers on their just knowing 'more' about a topic than other people. Psychologically, though, the difference is seen as due more to the type of knowledge that they have than the amount, with the **strategies** that they have for using it, and with the practice that they have had in dealing with problems in the area with which they are expert.

Chase and Simon (1973) studied expert chess players, to find out how their knowledge of chess differed from novices. One of their findings, confirming earlier work on memory, was that experts could easily recall the positions of pieces from a chess game that they had been shown only briefly, provided that the positions were actually taken from a game and so made sense to them. If the positions were random, then they were little better than the novices. This shows that they were able to make use of their knowledge of chess positions at the point of encoding, that is, they were able to see the positions that they were shown in terms of tactical chunks such as 'a king castled queenside' or a 'Sicilian defense' while novices would just see thirty-two separate pieces. Each position could therefore be represented as five or six chunks of pieces, enabling the whole board to be perceived in the brief time allowed, and reducing the amount of information that had to be remembered. The chunks are a form of **schematic knowledge** (see Topic D3) relating the appearance of the chess position to the moves and tactics that would have led to it, and its defensive or attacking purpose. The chunking also helps during problem solution, because it means that less load is placed on **working memory** (see Topic D2). Chase and Simon estimated that real chess experts knew around 50 000 chunks of information, which would have taken at least ten years of concentrated effort to acquire.

Domain specificity

The same general finding came from studies conducted on different 'domains' of knowledge such as within medicine or physics. Larkin *et al.* (1980) asked volunteers to sort a series of physics problems into different types, and found that experts tended to produce sets where the problems were related in terms of the physical laws or principles needed for their solution, whereas novices tended to sort them according to surface features in the description. The experts were able to classify and analyze the problems rapidly in terms of their special knowledge of the area. Their knowledge was not just more extensive, but it was organized in a different way and was better structured than that of the novices. Experts understood the interrelationships between apparently different parts of their knowledge. Lesgold (1988) asked people to talk aloud about how they were solving problems, and from these verbal protocols, found that experts spent a large proportion of their problem solution considering different ways that the problem could be represented, whereas novices concentrated on the most obvious representation. Once the experts had chosen a representation, however, it allowed them to solve the problem more rapidly than the novices.

The differences in encoding ability provided by structured knowledge allows experts to arrive at an appropriate representation of the problem faster than novices. It also allows them to attempt different solution strategies faster, because the problem-solving steps can also be chunked into routines. In effect, the experts have more procedural knowledge for problems in their own domain. While this helps them to rapidly solve problems that conform to their expertise and experience, it can also hinder their ability to solve novel or unconventional problems: here the **transfer** of knowledge from previous experience might prevent them being able to represent the problem in the way that supported solution (an example of the effect of **set** on problem solving; see Topic F1). They would be unable to 'see' the world from a novice point of view.

The domain specificity of knowledge means that experts are only experts in their particular area. Being an expert in one area does not mean that experts acquire any special problem-solving skills or strategies: they have no **meta-knowledge** about problem solving in general. Even changing the context of a problem can prevent expertise being applied. Carraher *et al.* (1985) found that Brazilian children who worked in a street market could calculate prices and give change very quickly, but when given the same sums to do in a school environment, were unable to solve them. Their expertise was in the form of buying and selling, and they did not see the relationship to other forms of mathematics.

Analogical reasoning

Domain specificity does not mean that experts can never transfer their knowledge to new situations. Paradoxically, human experts are sometimes rapidly able to solve problems that they have never experienced before, in which they apparently have no expertise. This is because they are able to use analogical reasoning, in which they recognize an abstract relationship between the novel problem and an old problem that they do know how to solve. For example, Gick and Holyoak (1980) asked people to solve an insight problem in which a surgeon must destroy an internal tumor by radiation therapy. The problem is to find how the surgeon can avoid destroying the healthy tissue through which the radiation must pass, and the answer is to use several low power beams from different angles, converging upon the tumor: each beam is too low power to kill tissue, and the critical level is only reached when they converge on the tumor. To look at the effect of analogical reasoning, they gave some people a similar, easier problem first. This problem involved a general who wanted to march his army towards a fortress that was only accessible by several small roads, each one of which was too narrow for the whole army to march down. If people were given the radiation problem alone, only 10% solved it. If people had solved the easier military problem first, 40% of them solved the radiation problem, and if they were also told that the military problem would help them solve the radiation problem, 80% solved it. The abstract relationship between these two problems lies in equating the fortress with the tumor, and the roads with the beams of radiation.

Reasoning by analogy is a common and ubiquitous feature of human thought, yet it is one that is extremely difficult to program a computer to do. To make an analogy between two problems, you have to identify the relationship that supports the analogy, and the attributes that the relationship applies to. If you are not told about the relevance of the earlier problem, you have to retrieve it from memory yourself. In the example above, there is no obvious connection between a military fortress and a tumor, but the context of the stories within which they are embedded makes them both things which are to be destroyed. Once this is noted, the way that the first problem was solved (by splitting the destroying force into small units) can be mapped onto the new problem.

Sternberg has argued that analogical reasoning is fundamental to human intelligence, and that by measuring the components of analogical reasoning it is possible to discover more about the sources of variance in intelligence. He identifies four components: **encoding** the terms of the analogy; **identifying** the relationship in the first part of the analogy; **mapping** this relationship onto the second part; and **application** of the mapping to produce the solution. His data suggests that people with higher intelligence scores spend more effort on the encoding component; this is equivalent to spending more time planning in problem-solving experiments. The richer the representation of the attributes of the items in the analogy, the more likely it is that a relationship can be discovered.

Theories of analogical reasoning have focused on the way that the mapping between analogical items is made. For example, Gentner and Gentner (1983) use a logical structure mapping approach in which entities within scenarios are represented as a hierarchy of propositional facts (e.g. in the military example: Roads are Small; Army is not Small; Army is divided into Small-Armies; Small Armies are Small; Small-Armies march down Roads; General destroys Fortress). When two scenarios have similar propositional structures, regardless of the content of each of the propositions, the structure mapping can be recognized and the analogy solved. Discovering the structure mapping is equivalent to the moment of insight in Gestalt conception of reasoning.

Expert systems   Research on the representation of knowledge by human experts helped in the development of artificial expert systems, especially in the area of diagnostic reasoning. Such expert systems simulate the experts' chunking of facts into higher-order units, and the strategies they have for discovering other pertinent facts in order to arrive at a conclusion. They are particularly useful when a course of action is required in a situation where there are a large number of facts to consider, and only a subset of them is relevant, because they can work through all relevant chunks of knowledge systematically. A well-known example is **MYCIN** (Shortliffe, 1976), which selects the appropriate antibiotic to prescribe for a bacterial infection. The **rules** in an expert system display some of the 'schematization' that an expert human does, by coming to intermediate conclusions that can be combined, rather than trying to leap directly to a conclusion. For example, the rules might take the form 'IF sore eyes AND runny nose THEN rhinoviral infection', 'IF upset stomach OR diarrhea THEN gastrointestinal infection' and 'IF rhinoviral infection AND NOT gastrointestinal infection THEN antibiotics not useful'.

Expert systems can also make use of a combination of problem-solving strategies. As well as the conventional **forward search**, as in the example above, they can make use of **backward searching** to search for new facts. An example might be a rule that says 'IF runny nose AND sore eyes THEN check for fever', which would cause rules that established the presence or absence of fever to be applied, such as 'IF forehead hot OR cheeks flushed OR thermometer reading over 38°C THEN patient has fever'. This saves time and effort because the expert system does not need to establish all possible facts before beginning to work towards a solution, but only needs to look for facts once they are found to be possibly relevant.

The most difficult stage in constructing an expert system is in obtaining the chunks of **schematic knowledge** from the human experts, a process known as **knowledge acquisition**. It is not productive to ask the human to write down all of their knowledge, because it is not stored in memory as formal, verbalizable 'IF...THEN...' rules. Instead a number of approaches must be taken, by asking the human about the domain, by using verbal protocols, by comparing the performance of different human experts on the same problems, and by a formal analysis of the domain. In consequence, the expert systems rarely behave in the same way that a human expert does, and while they can perform well in common situations, or in rare but well understood situations, they are less accurate in novel situations that were not considered during the knowledge acquisition process, or in situations which superficially resemble common ones but are 'exceptions to the rule'. For these situations special rules have to be written.

Creativity

Creativity is commonly seen as a genius or gift that individuals possess, but it is actually an ability that we all have, and is related to expertise. What makes particular individuals seem particularly creative in their field is partly their obsessive effort within that field, partly the expertise that they have built up through years of effort and application, and partly the environment within which they are working. Weisberg (1995) argues that creative thinking and insight is due to perseverance and expertise: creative people work long and hard studying the work of their predecessors and contemporaries, to become experts in their fields. What appears creative to others who are less expert is the result of applying ordinary problem-solving strategies to extraordinary knowledge. Csikszentmihalyi (1988) emphasizes that no one can be successfully creative on their own: they have to be in an environment which is both receptive to their ideas and stable enough to allow knowledge in a domain to accumulate. Gardner (1993) studied seven famous people, and found that they tended to:

- come from supportive but strict families;
- show early interest in their field;
- make creative 'breakthrough' only after gaining mastery in their field (after about 10 years);
- neglect or abandon personal relationships after their 'breakthrough'.

These characteristics are common to other classes of expert, and show the emphasis on concentration upon thinking about their area of creative expertise to the exclusion of other issues (see also Insight in Topic F1, and the effects of mood in Topic L3).

G1 DEDUCTIVE REASONING

Keynotes

Syllogisms	Syllogisms are a form of deductive logic, in which two facts (premises) are combined to reach a novel conclusion. The premises and conclusion can be in one of four moods: all A are B; no A are B; some A are B; and some A are not B, leading to 256 possible syllogisms – but only 24 are logically valid. People are very poor at recognizing which syllogisms are logically valid and which are not, leading some to argue that we do not actually think 'rationally'.
Content effects	Syllogisms that come to plausible conclusions, and use real, concrete terms, are easier to solve than those that use abstract symbols. We tend to accept as true any conclusions that match our prior knowledge and beliefs, even if the syllogisms are actually invalid. We may not be reasoning at all, but simply accumulating biased knowledge.
Atmosphere	People expect the conclusion to share the moods of the two premises. Where the premises differ in moods, a negative premise increases the acceptability of negative conclusions, and a particular premise (all A are B; no A are B) increases the acceptability of particular conclusions. This response bias may reflect the way people solve problems when under pressure to do so quickly, without reasoning.
Conversion	The moods 'no A are B' and 'some A are B' are bi-directional, and so can be converted into their reversed forms 'no B are A' and 'some B are A'. This is not the case for 'all A are B' and 'some A are not B', though, which do not imply 'all B are A' and 'some B are not A'. Many everyday facts can be converted, so if people assume that conversion is always allowed, they will make reasoning errors.
Misinterpretation	The logical meaning of **some** is 'not none', which can mean **all**. This is not the same as its everyday meaning. This can lead people to deduce that when they are told 'some A are B' that 'some A are not B', and vice versa, which are errors. People will accept conclusions that are possible consequences of the premises, even if they are not necessarily true, leading to errors.
Figural effect	In logic, the order of the premises and the order of terms in the conclusion should have no effect upon validity. However, syllogisms in the order A–B, B–C lead people to prefer an A–C conclusion, while the order C–B, B–A leads people to prefer a C–A conclusion. This cannot be explained by atmosphere, conversion or misinterpretation explanations.
Related topics	Mental models (G2) Inductive reasoning (G3)

Syllogisms   Whenever we decide that a conclusion follows from a set of facts that we know to be true, we are carrying out **deductive reasoning**. If our facts (the **premises**) are indeed true, and our deductive reasoning follows the laws of logic, then the conclusion must necessarily also be true. Nearly two and a half thousand years ago, the ancient Greek philosopher **Aristotle** realized that the simplest form of deductive reasoning occurs when two facts, each relating two terms (e.g. 'all dogs are animals' and 'all animals can breathe'), are combined to form a new, third fact (such as 'all dogs can breathe'). This form of logical argument is called a **syllogism**.

Each of the statements in a syllogism (the two premises and the conclusion) contain two of three terms: the **subject term**, S (here, 'dogs'); the **middle term**, M ('animals'); and the **predicate term**, P ('can breathe') and can be in one of the four **moods**, i.e. all A are B, no A are B, some A are B, some A are not B. The middle term that is shared between the two premises (M) is dropped in forming the conclusion, which thus relates the subject and predicate, S and P.

Since Aristotle's day, syllogisms have been traditionally organized into four **figures** (*Fig. 1*). The **major** premise contains the subject and middle term; the **minor** contains the predicate and middle term. The conclusion always links the subject with the predicate, eliminating the middle term. The two premises can be presented in any order.

Figure:	I	II	III	IV
Minor premise:	M–P	P–M	M–P	P–M
Major premise:	S–M	S–M	M–S	M–S
Conclusion:	S–P	S–P	S–P	S–P

Fig. 1.   The four Aristotelian figures of the syllogism. The Premises contain the Subject (S), Middle (M) and Predicate (P) term, and can occur in either order. The conclusion is found by eliminating the middle term, to link the Subject and Predicate.

The dogs/animals/can breathe example is organized in the first of these figures. Notice that, according to Aristotelian principles, the order of the major and minor premises is logically irrelevant: M–P followed by S–M is the same as S–M followed by M–P. Because each of the three statements can be in one of four moods, there are $4 \times 4 \times 4 = 64$ different syllogisms for each figure, and 256 different syllogisms in total.

Not all syllogisms are logically valid. For example, the syllogism 'all dogs are animals; some animals can swim; some dogs can swim' is not actually logically valid (even though you might believe the conclusion and know it to be true from your own experience). Compare it with the logically equivalent syllogism 'all dogs are animals; some animals can fly; some dogs can fly'. The conclusion clearly does not follow logically from the premises, and in fact, these two premises do not allow any logical conclusion to be drawn. Of the 256 syllogisms that can be constructed, only 24 are logically valid. Dickstein (1978) presented volunteers with all 64 pairs of premises, and asked them to choose which of the four logical conclusions followed, or whether they thought no logical conclusion followed. Overall, only 52% of their answers were correct (even choosing random answers would have given 20% correct).

Syllogisms have played a crucial role in the study of reasoning because they are the simplest possible form of logical reasoning, and yet we seem to be very poor at solving and understanding them. This raises the question of whether humans do reason logically, and indeed whether our thinking can be called reasoning at all: are we, in fact, irrational?

Content effects   Early attempts to explain our poor performance at syllogistic reasoning focused on the content of the statements, and in particular on the role of knowledge, believability, and the concreteness or abstractness of the premises. Wilkins (1928) found that people were better at solving syllogisms that used real, concrete terms (such as the dogs/animals examples) than abstract symbols (letters). However, she also found very strong misleading effects of the believability of the conclusion: if the conclusion concurred with her volunteers' prior knowledge and their beliefs about the world, they were more likely to accept it as valid regardless of the logical validity of the syllogism that led to it. In contrast, there was a tendency not to accept logically valid syllogisms with implausible conclusions (i.e. those where the conclusion logically followed from the premises, but where the premises were not actually true). This finding was quantified by Evans, Barston and Pollard (1983) (*Fig. 2*). The implication of this for our understanding of the processes by which we reason are disturbing, for if we accept things as true simply because they are consistent with what we believe, it is likely that our beliefs are themselves derived from similarly invalid reasoning. We may not be reasoning at all, but simply accumulating biased knowledge based upon consistency with experience (see Confirmation bias in Topic G3).

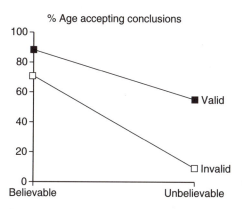

Fig. 2.   People are more likely to accept a believable but invalid conclusion than an unbelievable but valid conclusion (from Evans, et al. (1983).)

Atmosphere   A consistent finding in research with syllogisms is that people have greatest difficulty with premises and conclusions that involve the word **some**, and especially with the mood '**some A are not B**'. Woodworth and Sells (1935) suggested that the presence of moods in the premises might predispose people to expect the conclusion to take a similar form, by creating an **atmosphere.** They pointed out that the moods could be seen as making either positive or negative claims, and either universal or particular claims (*Fig. 3*). If both premises are universal, then a universal conclusion is preferred, but if both are particular, then a particular conclusion in preferred. Where the premises differ in moods, a **negative premise** increases the acceptability of negative conclusions, even when the other premise is positive; and a **particular premise** increases the acceptability of particular conclusions, even when the other premise is universal.

	Universal	**Particular**
Positive	All A are B	Some A are B
Negative	No A are B	Some A are not B

Fig. 3.   The atmosphere of premises may lead people to expect the conclusion to take a similar form. Where the premises are mixed, the negative and particular moods are preferred in the conclusion.

The atmosphere hypothesis is essentially about **response bias**, since it says nothing about the reasoning process itself. However, it may correctly reflect the way that people choose between different alternatives when they are under pressure to do so quickly (as is commonly the case in reasoning tests).

Conversion

The conversion hypothesis was an attempt to replace the atmosphere hypothesis by an account that explained reasoning errors in terms of processing rather than response bias. Chapman and Chapman (1959) noted that one of the difficulties in relating everyday reasoning and syllogisms is that many everyday statements are true regardless of the order of the terms, whereas logical statements are not. The moods 'no A are B' and 'some A are B' can be **converted** into their reversed form, e.g. the fact 'no trees are animals' does imply that 'no animals are trees', and 'some dogs are pets' does imply that 'some pets are dogs'. This is not the case for 'all A are B' and 'some A are not B', though. While everyday, definitional statements such as 'all boys are young males' can be converted to 'all young males are boys', statements about class membership such as 'all boys are humans' cannot be converted to imply that 'all humans are boys'. Similarly, 'some animals are not dogs' cannot be converted to imply that 'some dogs are not animals', no matter what their owners might think. Clearly, if people routinely assume that all premises can be converted, this will lead to errors in reasoning. Although for many syllogisms the two hypotheses make the same predictions, in a small set of cases they differ, and Chapman and Chapman's data fitted the predictions of the conversion hypothesis better than the atmosphere hypothesis.

Misinterpretation

The conversion hypothesis rests on a difference between the everyday use of language and its strictly logical meaning. While 'all' and 'not' mean the same in logic as in everyday use, 'some' has a particular meaning in logic but is rather flexible in everyday use. The difficulty with understanding 'some' in logical problems is that it really means 'not none'. In common use 'some' has a meaning that corresponds to a smallish number, not as many as 'a lot', but more than 'a few'. It certainly does not mean 'all', whereas 'not none' clearly does include that possibility. However, trying to think about premises such as 'not none animals are not dogs' makes it clear why 'some' has been used instead. Its logical definition to include 'all', though, leads to reasoning problems, as the following example shows.

The common interpretation of 'some acrobats are bankers' suggests that there are many acrobats, and that at least one of them is a banker, but that others may not be. You would probably assume that the person telling you 'some acrobats are bankers' has reason to know that some of the acrobats are actually not bankers, or otherwise they would tell you 'all acrobats are bankers'. So, you reason, 'some acrobats are not bankers', and you have made an error. In logic, 'some acrobats are bankers' really means that we know that at least one acrobat is a banker but we have not checked the rest of them, they might all be bankers: it explicitly does not mean that we know that some acrobats are not bankers. The same confusion occurs with 'some acrobats are not bankers'. It seems to imply that the person giving us the information knows that 'some acrobats are bankers', or else they would tell us that 'no acrobats are bankers'.

Grice's (1975) maxims about communication (see Topic H2) say that a speaker will be informative and not withhold relevant information. The logical meanings of 'some A are B' and 'some A are not B' conflict with these **Gricean maxims**, whereas the Gricean understanding of them allows misunderstanding of the logical meanings to occur.

A similar argument has been applied to our understanding of what it means for a conclusion to be logically valid. For a conclusion to be valid, it must not only be possible, it must be necessarily true. That is, while it is possible that 'some A are B' and 'some A are not B' can both be true at the same time, it is not necessarily inevitable that they must both be true at the same time. A logical conclusion is only valid if it must be true, given its premises. Dickstein (1981) argues that people will accept conclusions that are possible consequences of the premises, even if they are not necessarily true. This is known as the **misinterpreted necessity** argument.

Figural effect

Johnson-Laird threw doubt on the conversion and atmosphere hypotheses by reporting systematic differences in acceptability for some syllogisms that ought to have been equally acceptable if people had been using conversion or atmosphere. He noted that when shown the syllogism 'all B are A; all C are B', people accepted the correct 'all C are A' conclusion twice as often as its incorrect conversion 'all A are C'.

Johnson-Laird and Steedman (1978) went on to show that in general when given premises that included the terms in the order M–P, S–M (i.e. the premises of Aristotle's Figure I in *Fig. 1*), people preferred conclusions of an S–P form, whereas when they saw premises with terms in the order P–M, M–S (i.e. Aristotle's Figure IV in *Fig. 1*), they preferred conclusions of a P–S form. For the other two figures, there was no such bias. They called this preference for the form of the conclusion the **figural effect**.

Note that according to traditional forms of logic, a conclusion must be S–P, not P–S. Traditional logic starts with a conclusion and works back to identify the premises, so it does not make sense to talk of a P–S conclusion; the first term in the conclusion is always the Subject (S) and the second is always the Predicate (P). One premise will contain S, and the other P, but it should not logically matter what order they are in. By giving people the premises and asking them about the conclusions, researchers had inadvertently made the order of presentation relevant, and the discovery of the figural effect suggested that however people were solving these logical problems, it was not by applying rules of logic.

G2 MENTAL MODELS

Keynotes

Mental models	Johnson-Laird and colleagues argue that reasoning is based upon mental models of premises. People construct a mental model of the first premise, then add in the information from the second premise to construct as many combined models as they can, and finally frame a conclusion to express any relationship that they can find that holds in all possible models. This account attributes reasoning errors to limited memory capacity and biases in the motivation to consider potential combinations.
Mental logics	Mental logic theories of reasoning hold that we extract abstract symbolic or propositional representations from semantic facts, and use inference rules to combine and evaluate these propositions. We acquire the ability to carry out propositional logic as we mature. Since we develop rules through experience with the world, problems whose structure match frequently-occurring situations should be easier to solve than unusual problems.
Models or logics?	Both approaches explain problem difficulty in terms of limited capacity working memory, and make similar predictions, so it is difficult to distinguish them empirically. Content effects may favor the models approach, since once problems have been represented symbolically as propositions, it should not matter what their content is.
Related topics	Representations (E1) Inductive reasoning (G3) Gestalt problem solving (F1)

Mental models   Since the 1970s, Johnson-Laird and his colleagues (e.g. Johnson-Laird, 1983) have been advancing an account of reasoning that is based upon the generation of mental representations or models of the world. In syllogistic reasoning (see Topic G1), the presentation of a premise allows a person to construct a mental model of the relationship between the two terms. The second premise allows another model to be generated. Because the two models share a common term, they can then potentially be combined to form a number of possible alternative models. If there is no way of combining the models, or if there are several possible ways of combining the models, then the premises do not lead to any single conclusion. If they can only be combined in one way, then that combined model can be interpreted to draw a logically valid conclusion. Mental models theory has also been influential in research into the inferences that must be made during language comprehension (see Topic I2).

Figure 1 shows how two models generated from the premises 'no acrobats are bankers' and 'some cowboys are bankers' can be combined to reach a conclusion. Each row in the model represents a different case, so the six rows in the first model represent three acrobats (none of whom are bankers) and three

1st premise	+	2nd premise	→	Conclusion
'none of the acrobats are bankers' could be represented by a mental model of acrobats and bankers:		'some cowboys are bankers' could be represented in a similar way:		If these two premises are given together, their mental models could be combined:

1st premise	2nd premise	Conclusion

acrobat
acrobat
acrobat
banker
banker
banker
	banker = cowboy
	banker = cowboy
	(banker)
	      (cowboy)
		acrobat
		acrobat
		acrobat
		    banker = cowboy
		    banker = cowboy
		    (banker)
		       (cowboy)

Each 'line' is an instance of a different logical entity in the model.	The brackets indicate uncertainty about the existence of a condition – the banker who is not a cowboy, and vice versa, may or may not exist	Having formed this combined model, it would be tempting to conclude that 'none of the acrobats are cowboys', or that 'none of the cowboys are acrobats'.

However, there are other ways of combining the two premises:

acrobat
acrobat
acrobat =   cowboy
    banker = cowboy
    banker = cowboy
    (banker)
	acrobat = cowboy
	acrobat = cowboy
	acrobat = cowboy
	    banker = cowboy
	    banker = cowboy
	    (banker)

These two models lead to different conclusions:

'some acrobats are not cowboys'   **'all acrobats are cowboys'**
'some cowboys are not acrobats'   **'some cowboys are not acrobats'**

Fig. 1.   The mental models of the two premises can be combined in three different ways, each leading to several conclusions, but only one conclusion is consistent with all possible combined models. 'Some cowboys are not acrobats' is therefore the valid conclusion from these two premises.

bankers (none of whom are acrobats). This is the only way in which this premise can be modeled. The four rows of the second model represent two bankers who are also cowboys, one banker who is not a cowboy, and one cowboy who is not a banker. This is not the only way in which this premise can be modeled, because we do not know for sure about the existence of either the last two rows. A combination of these two models can be made simply by overlapping the three bankers from each model. The resulting model has seven rows, and can now be inspected to see what can be deduced about the relationship between acrobat and cowboy. It might be tempting to deduce that 'no acrobats are cowboys' or 'no cowboys are acrobats', which are certainly true for this model. The model is also consistent with the conclusions 'some acrobats are not cowboys' and 'some cowboys are not acrobats'. However, this is not the only way that the two models can be combined. The rest of *Fig. 1* shows two slightly different combinations, one in which the cowboy who was not a banker turns out to be an acrobat, and one in which all of the acrobats turn out to be cowboys. All three models lead to a range of different conclusions, but only one of the conclusions can be drawn from all combinations of models. Whatever combined model we choose, 'some cowboys are not acrobats' is always consistent, and so this must be the valid conclusion for these two premises.

Johnson-Laird's theory proposes that people i) construct a mental model of the first premise; then ii) add in the information from the second premise to construct a combined model (taking into account as many different ways that this can be done as they can manage); and iii) finally frame a conclusion to express any relationship that they can find between the 'end terms' that holds in all models.

It is the second step – the **combining** of premises and the **juggling** of different possible models – that gives rise to error. Some syllogisms require one model to be built, others require two or even (as here) three models to be built and examined. The more models that are required, the more likely it is that one will be overlooked or forgotten. Johnson-Laird's theory copes well with the **figural effect**, since the order of the A, B and C terms within the premises affects the number of operations that are required to construct the model. With the A–B, B–C order of the first figure, it is easy to integrate the second premise with the first, biasing the reasoner to an A–C conclusion. With the other figures, one or both of the premises has to be 'rotated' (note that this is *not* the same as conversion, see Topic G1), or the second premise has to be modeled first. The greater the number of operations, the harder the syllogism is to solve.

The mental representations at the heart of this account of reasoning are analogical, in that they bear some structural relationship to the real world. Although it is natural to think of mental models as being visual in nature, Johnson-Laird emphasizes that this is not necessarily so; they can be in some abstract representational form. However, their analogical nature means that they may be inaccurate, and even if they are accurate, errors in reasoning may occur due to limited working memory capacity (see Topic D2), since all potential models must be constructed and compared to discover which conclusions are consistent across them. Believability has its effect by increasing the motivation of reasoners to continue constructing models until they find one that allows a believable conclusion to be deduced, while they will be less motivated to build models to justify implausible conclusions.

Mental logics   A group of related theories of reasoning do not rely on analogical mental representations of the world. In contrast, they hold that we extract abstract symbolic or propositional representations from semantic facts, and use inference rules to combine and evaluate these propositions (see Topic E1). In essence, they argue that our minds acquire the ability to carry out propositional logic as we mature, and so the theories can be grouped together under the headings of mental logics or mental rules. Piaget's theory of development is an example (see Topic J1): when a child passes from the concrete to the formal operational stage, they are learning to use propositional logic rather than analogical representations. Braine and colleagues have argued that the difficulty of reasoning problems depends upon the reasoner possessing inference rules that match the structure of the propositions they are being asked to deal with. Since we develop rules through experience with the world, problems whose structure match frequently-occurring situations should be easier to solve than unusual problems. Only individuals with well-developed mental logics will be able to solve unusual problems. Using their own estimates of which rules are 'difficult' to acquire, Braine *et al.*'s theory was able to predict reasoning errors quite well (a correlation of 0.73, where 1.00 would be perfect prediction and 0 no prediction).

Rips (1994) proposed a 'natural deduction' approach called PSYCOP which is similar to Braine's theory, but which as well as applying rules to work forward from propositions to a conclusion, has some rules which can work backwards from a desired conclusion towards the known propositions. This allows it to work with fewer inference rules. In PSYCOP, the likelihood of a rule being chosen depends upon its availability, which will vary according to the individual's state of expertise as a reasoner and the match between the problem context and the rule's form. Using estimates of availability derived from human performance on syllogistic reasoning tasks, PSYCOP was able to predict errors with a correlation of 0.88.

All mental logic approaches allow five classes of errors in reasoning: encoding errors, where the reasoner incorrectly forms the propositional information from the known facts; production errors, where they make a mistake converting the propositional result into real-world context; logical errors, where they have an inappropriate inference rule or have not yet acquired a rule that is required; heuristic errors, where they have the appropriate rules but make a mistake in choosing which to apply; and capacity errors, when the complexity of processing required to apply the rules exceeds the limits of working memory.

Models or logics?   It has proven difficult to empirically distinguish between the analogical mental models approach and the propositional mental logic approaches, because both types make similar predictions about the difficulty of reasoning problems. Both also explain problem difficulty in terms of the load on a limited capacity working memory. The content effects are perhaps the strongest evidence in favor of analogical models, since once problems have been represented symbolically as propositions, it should not matter what their content is: it is difficult to see how mental logic approaches can deal with the effects of believability without compromising their central strength.

G3 INDUCTIVE REASONING

Keynotes

Confirmation bias	People prefer to seek evidence that confirms their hypotheses, but however much evidence confirms a hypothesis, only one piece of contradictory evidence is needed to disprove it (falsify it), so it is better to seek falsifying instances.
Selection problem	A task in which people have to turn over any of four cards (**P**, **Q**, **not P** and **not Q**) to seek evidence about the validity of a rule of the form 'if **P** then **Q**'. They tend to examine the cards that confirm the rule (**P** and **Q**) rather than the card that can potentially falsify it (**P** and **not Q**).
Schematic reasoning	Instead of reasoning abstractly in the selection task, people may be making use of ways of reasoning that they have learnt or which are innate, and which deal with everyday social contexts in which costs and benefits are traded off.
2–4–6 task	A task in which people have to generate tests of a hypothesis about a rule for the generation of triplets of numbers. They tend to generate tests that they expect to confirm their hypothesis instead of falsifying it, so never discover that they are incorrect. It has been argued that this is in fact a positive test strategy rather than a confirmation bias.
Related topics	Deductive reasoning (G1) Decision making (G4)

Confirmation bias Given some facts, reasoning can either deduce a more specific fact that is their logical consequence, or can induce a more general fact that is consistent with the existing facts, but which may or may not be true. Scientific reasoning operates through a sequence of induction and deduction: observations of the world are used to gather facts, and then inductive reasoning is used to generate general theories, from which specific hypotheses are then obtained by deductive reasoning. According to Popper, scientific reasoning can only operate if hypotheses are both testable and falsifiable.

A common finding in psychology and philosophy has been that people are not very good at generating testable hypotheses, nor at designing tests that can falsify hypotheses. Instead, they seem to try to confirm their hypotheses. This is the wrong approach, because however many confirming facts you find, you only need one falsification to disprove a theory (this is why science can never prove theories; see Topic A2). Wason devised two classic reasoning problems that have been extensively used in research on this **confirmation bias**: the selection problem (Wason, 1966) and the 2–4–6 problem (Wason, 1960).

Selection problem In the selection problem (*Fig. 1*), people are told a rule of the general form 'if **P** then **Q**', and asked to discover whether the rule is true or not by testing pieces of evidence. In scientific terms, the rule is a hypothesis, and the nature of the

These cards all have a letter on one side and a number on the other: | E | | K | | 4 | | 7 |
RULE: 'if the letter is a vowel, the number is even'
Which cards would you turn over to test this rule?

Whenever I go to York, I go by car | York | | Leeds | | car | | train |

I only eat spaghetti off blue plates | spaghetti | | Blue plate | | lasagne | | Red plate |

Fig. 1. Three versions of Wason's Selection Problem, with its original abstract form at the top. The middle, concrete version is easier to solve, but the bottom version is still difficult, even though it also uses concrete material.

tests they carry out reveals that they actually seek to confirm the rule (by finding evidence that matches it, such as cases where '**P and Q**' are true) rather than trying to falsify it (by finding evidence that is inconsistent, such as a case where '**P, not Q**' is true).

In its classic form, the problem involves four cards, each bearing a letter on one side and a number on the other. The rule to be tested is 'If a card has a vowel on one side, then it has an even number on the other side'. The cards are arranged so that two display the letters E and K, while two display the numbers 4 and 7. To falsify this rule, people need to inspect the other sides of the cards bearing the **P** term (the vowel, E) and the **not Q** term (the odd number, 7). If E has an odd number on the other side, or if 7 has a vowel, then the rule has been falsified. Any other outcomes either confirm the rule (such as E having an even number, or 4 having a vowel) or are irrelevant (such as 4 having a consonant, or K having an odd number). In practice, people do tend to turn over the **P** card (E), which is correct, but also the **Q** card (4), which is incorrect, since it cannot help them falsify the rule.

Schematic reasoning

At face value, this suggests that people do not naturally reason scientifically. Many explanations have been proposed for this pattern of findings. Evan's (1984) **matching bias** is the simplest, because it argues that people just pick cards which match terms given in the rule. While this accounts for much of the data, variations that manipulate the content of the rule and the social setting have shown contrasting patterns of performance with structurally identical rules, and have led to explanations based on experience or schematic reasoning. Johnson-Laird *et al.*'s **memory cueing** hypothesis argues that people perform better when the selection task uses concrete terms and rules that they are familiar with, such as 'if a person is drinking alcohol, they must be over 18' (at least in some parts of the world!)). Cheng and Holyoak's (1985) **pragmatic reasoning schemas** are abstract versions of the memory cueing hypothesis generalized to situations such as those involving permission (to do X, you must Y), obligation (if X, you must Y) and denial (if X, you cannot Y). Cosmides' (1989) theory takes an evolutionary perspective, arguing that our species has innate mechanisms to detect violations of the **social contract**, whereby people gain a benefit to which they are not entitled.

Manktelow and Over (1991) have suggested that the crucial aspect of selection tasks is their **deontic** nature – they involve both **permissions** (allowing an action) and **obligations** (requiring an action). Although familiar content is not necessary for deontic reasoning, an explicit rationale for the rule (either through experience, or in the instructions) facilitates accurate performance. With a rule such as a parent saying to a child 'if you tidy your room, you can go out to play', people check the 'tidy room' and 'no play' cards when checking whether the parent has broken the rule by reneging on their permissions, but the 'untidy room' and 'play' cards when checking whether the child has fulfilled their obligations. Both of these sets of choices are completely **rational** in the context of a deontic social situation, although one is **irrational** in the original terms of the selection task, namely scientific hypothesis-testing. Apparently confirmatory biases on the original task may therefore owe more to the practical aspects of human reasoning than to the theoretical aspects of the task.

2–4–6 task

This task was developed as a direct analogy of scientific reasoning. People are told that there is a rule that generates triplets of numbers, and that '2–4–6' is an example of a triplet generated by the rule. They are asked to find out what the rule might be, and that they can get more evidence to help them by generating triplets, which the experimenter will classify as either fitting or not fitting the rule. The typical pattern of results is that people come up with a hypothesis such as 'ascending even numbers', and then generate triplets that would fit this hypothesis – they seek confirming evidence, rather than trying to falsify their hypothesis by generating triplets that do not fit it. Because they only test confirming evidence, they are surprised to find that their hypothesis about the rule is not correct (in fact, it is 'ascending numbers', and so odd numbers can be included too). This task has been cited as further evidence of a **confirmation bias** in human reasoning.

Klayman and Ha (1987) noted that the real rule was more general than the volunteers' typical hypotheses. When they used a real rule that was more specific ('ascending, consecutive, even numbers'), then although people still started off in the same way, by generating triplets to fit their initial hypothesis, the answers they got were now sometimes that the triplets did not fit the rule (e.g. the triplet '2, 10, 100'). They were then able to refine their hypotheses in the appropriate direction, towards the more specific rule, and discovered it quite quickly. This indicates that people are able to use falsifying evidence, when they get it. Klayman and Ha argue that people do not seek confirming evidence, but have a **positive test strategy**. When searching for a very specific rule, or a rare set of circumstances, starting with a more general rule (or set of circumstances) and following a positive test strategy does allow both confirming and falsifying evidence to be collected. Confirming evidence tells you that you are working towards discovering the true rule, while the falsifying evidence helps you identify the ways in which your current hypothesis is too general. Klayman and Ha argue also that this is the more typical way in which people go about exploring the world, and that it is more representative of the way that scientific discovery operates. It is only when an artificial and very general rule needs to be discovered (as in the 2–4–6 task) that this natural strategy fails to provide falsifying evidence.

G4 DECISION MAKING

Keynotes		

Keynotes

Normative models Economic models of behavior assume that we take into account all relevant evidence to ensure that we come up with the optimal, or rational, decision, given the evidence that we have at our disposal. These models do not match actual behavior.

Probability In predicting sequences of events, we tend to make use of probability matching, in which our predicted sequences look like typical sequences, instead of predicting runs of the most probable event, which do not look like typical sequences. To estimate probability accurately, we need to have accurate memory for the frequency of previous events. We actually overestimate the frequencies of rare events and familiar events.

Heuristics Rather than evaluating costs and benefits or calculating probabilities, we save time and effort by using heuristics that are rapid, economical, and reasonably likely to work. Tversky and Kahneman describe three heuristics that can be drawn on to account for many of the biases found in probabilistic reasoning: availability, representativeness, and anchoring and adjustment.

Subjective utility The utility of a gain or a loss is subjective, and depends on the difference that it will make to the reasoner's *status quo*: a small gain is worth more to someone who has little than to someone who has a lot.

Framing Phrasing mathematically equivalent choices as situations involving a loss or a gain can alter the choices people make in ways that do not accord with normative theory. Normally people are loss averse. When choosing between a situation resulting in a sure gain and a situation with a chance of either a larger gain or no gain at all, people are risk averse and seek the sure gain. When choosing between a sure loss situation and a situation with chance of a larger loss or no loss, people are risk seeking, and take the chance of the large loss.

Related topics Concepts (E3) Inductive reasoning (G3)

Normative models

When making a decision such as 'should I take a coat with me or should I buy an umbrella?' we are making a complex set of judgments about the probability of future events (i.e. whether or not it will rain) and the costs and benefits to us of alternatives (i.e. the cost of the effort carrying around a coat all day, or of spending some money on an umbrella, and the benefit of staying dry and comfortable if it does rain). Normative models predict what people ought to do to maximize their personal gain, i.e. maximize their benefits and minimize their costs. They have been developed to help decision makers take into account all of the relevant evidence to ensure that they come up with the optimal, or rational, decision, given the evidence that they have at their disposal, and these form the basis of models of economic behavior. However, they assume that we do in fact act rationally, and do take into account all appropriate evidence accurately. This does not correspond to our actual behavior.

Probability

There are several ways of defining the probability of a future event. A frequency-based model asks 'If I gave this event 100 chances to occur, how many times would it really happen?', and is good for events that happen a lot, and whose history is known. A logical model asks 'Out of all possible ways that things can happen, how many of them will result in this event?', and is good for rarer events, which can be broken down into causal chains. A Bayesian model asks 'Given the association that I know between this event and other events, what other events have occurred recently?', and is good for events that are causally related or highly associated to other events. It seems that people cannot follow any of these normative models, or if they can, they do not base their behavior on them.

Imagine that you are shown an opaque jar containing 1000 small colored balls, and you draw out 10 balls, eight of which are red and two of which are white. You can probably estimate the probability that the next ball you pick will be red (it is 0.8, based on your sample). If asked to predict the color of the next ball that you will pick, you would be sensible to say 'red', since it is four times as likely to be red as white, on the frequency based evidence at your disposal. However, if asked to predict the color of the next fifty balls, in order, what would you say?

The normative answer is to say 'red, red, red...' for all 50 balls, since each ball is more likely to be red than white, and you would be right 0.8 of the time. A more human response is to follow a strategy of **probability matching** and to say something like 'red, red, red, white, red, white...' producing a sequence that is 80% red and 20% white. While it is true that this sequence is more typical of the actual type of sequence, you will be 'correct' for far fewer balls within it, and so be less accurate overall. Each time you say 'white', after all, you only have a 20% chance of being correct, instead of 80%. It is as if you are trying to follow a strategy that will lead you to be right every time, rather than the most optimal (but also imperfect) strategy.

There is also a tendency for people to overestimate the frequency of low probability (or surprising) events and familiar (or easily recognized) events. Kahneman and Tversky (1973) read out a list of 39 names, with 19 famous men and 20 unknown women (or vice versa), and then asked people if they had heard more male or female names. People incorrectly said that the gender that was most common in the list was whichever one had been represented by famous names. In these cases, our memory is not allowing us to access accurate information about frequencies, and so we cannot follow any normative model based on the frequency of past events – which in practice rules out all three models of probability.

Heuristics

In estimating frequencies, we have to rely on our memory for past events – we do not have accurate frequency counts to rely on. Rather than spend time racking our brains to come up with every instance we can think of, or to calculate logical odds, we save time and effort by using methods that are rapid, economical, and reasonably likely to work: **heuristics**.

Tversky and Kahneman (1974) describe three heuristics that can be drawn on to account for many of the biases found in probabilistic reasoning: **availability, representativeness,** and **anchoring and adjustment.** The easier it is to recall an example of an event (i.e. the more available an example is in memory), the greater the probability that is assigned to it. Air disasters are widely reported, and so it is easy to recall an example of a crash, whereas flights that do not crash are much more common, so do not get reported, and are less salient. Road accidents happen frequently, but are cleared up fairly quickly and are not widely reported, so we rarely see them – and so our estimation of their frequency is reduced. The consequence is that most people believe (incorrectly) that it is safer to travel by road than air. Availability can explain the familiarity effect: the 'famous' names were easier for people to recall, and so the frequency of whichever gender was paired with fame was overestimated.

Representativeness means that an event is judged as an instance of a class whose stereotypical members it resembles, regardless of other information. Imagine a woman called Linda, who is 31 years old, single, outspoken and very bright. She studied philosophy, and as a student was very concerned with issues of discrimination and social justice, and was a member of several political societies. Which of the following statements about Linda is more probable: (a) Linda works in a bank; (b) Linda works in a bank and buys organic food? Given what you know about Linda, it seems as if (b) is likely to be true, and you may be tempted to choose it. Whenever (b) is true, though, (a) is also true, so the probability of (a) must be at least equal to (b), and it is probably greater. So (a) is the correct choice. If you choose (b) you are doing so because that option seems more representative of the sort of person you think Linda is. You are not doing so on a normative model of probability. Representativeness is also seen in the famous **gamblers' fallacy**, also known as the 'law of averages', which says that if an event has occurred less often than its probability suggests, then it is more likely to happen in the near future. The probability-matching choice of a typical sequence of balls drawn from a jar can be explained by this heuristic.

The third heuristic, anchoring and adjustment, suggest that once people have made an initial estimate of a probability (their **anchor**), they start to **adjust** it according to other evidence. The general trend is to make insufficient adjustment, and so the anchor that people are given (or which is most **available** or most **representative**) biases their final answer. If you are given five seconds to estimate the answer to the sum $1 \times 2 \times 3 \times 4 \times 5 \times 6 \times 7 \times 8$, you will start by multiplying the small numbers, then give up and guess at a value of about 500–1000. If, instead, you are asked to estimate $8 \times 7 \times 6 \times 5 \times 4 \times 3 \times 2 \times 1$, you will start by multiplying the large numbers, realize how quickly the total is growing, and guess at a total of maybe 3 000 to 5 000. The anchor in this case is the total you had got to by multiplying the first few numbers, and you then adjusted it somewhat to allow for the numbers you had not got to. The two orders change the size of the initial anchor, and so lead to different estimates. In either case, your subsequent adjustment was probably insufficient, however, since the correct answer is 40 320.

Subjective utility   In assessing the costs and benefits of different courses of action, the *status quo* of the reasoner has to be taken into account. This is because of the 'law of diminishing marginal utility', which states that the more you have of something, the less you value even more of it: a small amount of money is of more value to a poor person than to a rich person. You will consequently stop taking an action to get benefits once you have reached a certain level of benefit, even if the cost:benefit ratio of the action remains constant. The utility of a gain or a loss is subjective, and depends on the difference that it will make to the reasoner's *status quo*.

Other possible ways of obtaining the benefit can also influence the subjective utility of a benefit: the chance of winning a state lottery may be minuscule, meaning that the expected benefit (say €3 000 000 divided by 15 000 000 tickets, or 20 cents) is less than the certain cost of the ticket (€1). However, there is probably no other way in which most of us could ever acquire €3 000 000, and there are many ways in which we can regain one lost euro, so the very remote chance of an enormous and unique benefit outweighs a small loss.

Framing

Subjective utility leads to a phenomenon known as **loss aversion**, whereby a possible loss of some amount is more aversive to people than a possible gain of the same amount is attractive, resulting in them choosing the *status quo*. However, Tversky and Kahneman (1981) found that in certain circumstances people would also show **risk aversion** and **risk seeking**. If you could either open one box that certainly contains €500, or another that has an 85% chance of holding €600, but a 15% chance of being empty, which would you choose? Most people go for the certainty of €500, even though the other alternative has a higher expected value of €510 (€600 × 0.85). It seems as if we adjust our *status quo* to put ourselves in the position of already having obtained the €500. Choosing the other box can now only gain us €100, but may lose us €500. Reversing the problem, would you rather press a button that will deduct €500 from your bank account, or press another button that will deduct €600 85% of the time, but do nothing 15% of the time? In this terrible situation, most people opt to take the risk and press the second button. Again, by taking the certain loss of €500 into account, pressing the other button appears to offer us the 15% of gaining this large amount back, for the small cost of an 85% chance of losing a further €100. **Framing** the options as a gain or a loss alters the choice people make in ways that do not fit with normative models.

H1 COMMUNICATION

Keynotes

Nature of language	Language involves all of our cognitive abilities, and its psychological aspects are the topic of psycholinguistics. Languages enable communication, differ between cultures, are primarily vocal but use a subset of all possible vocalizations. They consist of arbitrary units that can be arranged according to grammatical rules to create an infinity of novel utterances, that need not be true now or ever. An individual speaker will modify the **register** of their language depending upon the situation, and while they know what is grammatical and what is not (their **competence**) they do not always speak grammatically, nor use all of the words they understand (their **performance**).
Transformational grammar	Chomsky argued that instead of being conditioned, language was supported by an innate language acquisition device that gave the ability to acquire vocabulary and to learn grammatical rules. He proposed that all languages shared a deep structure and a set of phrase structure rules that could be used to produce the surface structure of utterances. All sentences consist of a noun phrase and a verb phrase, which in turn contains a noun phrase. The noun phrase can itself contain subsidiary sentences, or can be a simple combination of article, adjective and noun. Chomsky's approach has dominated modern psycholinguistics.
Modularity	Language is determined by genetics (because only humans develop it) and also by environment (because we speak the language we hear around us when we are babies). Chomsky's arguments for the innateness of language suggest to some that language ability is modular, that it has a dedicated set of cognitive processes that operate upon linguistic input to produce meaning, and that it is localized in the brain. Others argue that it depends a great deal on other cognitive abilities such as memory and inference. Debate continues about whether language development is continuous or whether it passes through discrete stages; and whether there are optimal or critical periods during which exposure to language is necessary for neural development.
Primate language	Although closely related primate species are not physiologically able to speak, they may share rudimentary language-acquisition abilities. Attempts to raise chimpanzees in linguistically rich environments have failed to show much evidence for language production. Chimps fail to learn any grammatical structure, and remain at the level of a two-year-old human speaker. Their ability to comprehend language is much greater, though, and so in humans the ability to produce language might have developed following sound comprehension skills initially acquired for other purposes.
Related topics	Discourse (H2) Language and thought (I3) Understanding words (I1)

Nature of language

Human language use is a unique and characteristic attribute of our species' mental behavior. No other species' communications approach the complexity and flexibility of human language; language pervades our thoughts and mental processes so thoroughly that it is difficult to study any aspect of cognition without involving language in some way. Language also requires every aspect of our cognitive capabilities. To speak and to listen, and to read and write, we need to be able to perceive linguistic units; to remember what each unit means; to infer the meaning of the group of units, taking into account what has gone before and the context of the communication; we need to be able to turn our thoughts into linguistic structures, and to produce language units that other people can understand. The scientific discipline of linguistics is devoted to the study of human language; where it overlaps with psychological questions and theory the field is called psycholinguistics. Work in this area over the past 40 years has helped us to understand many aspects of language use, and language disorders, as well as showing the contribution to language use of cognitive functions such as short-term memory (see Topic D2), long-term memory (see Topic D3) and attention (see Section C).

Defining language precisely is not easy, but the following attributes have been proposed:

- Language enables communication between individuals.
- Language is culturally transmitted and varies across cultural groups.
- Language uses primarily vocal sounds, but only a subset of all possible vocal sounds.
- Language units are arbitrary symbols that need not have any correspondence to the things they represent.
- Language has a grammatical structure that can be analyzed on many levels.
- Language units can be arranged according to this grammar to produce novel utterances and to convey novel ideas.
- The ideas need not currently be true, and might never have been or never be true.

Hockett (e.g. 1966) attempted to develop a list of such linguistic universals that were true across all human languages, but which did not all apply to forms of communication used by other species, such as bird song, or the 'waggle dance' of the honey bee (see Topic I3).

A person's language is not a single thing. A person will speak in different ways to different people, and in different settings. Consider the way that you would speak to a young child, or to a friend, or to an official of some type. Consider the way that you communicate to a relative face-to-face, compared to speaking to them over the phone, or when writing a letter to them. You might be using the same language, but in a different **register**. You can also hear or read and comprehend many more words than you use in your everyday speech or writing; and you can also understand the meaning of grammatical constructions – and misconstructions – that you would not generate yourself. This indicates a distinction between language **competence** and language **performance**. The former always exceeds the latter, by definition, and it is the former that is usually studied within psycholinguistics. Actual speech is extremely fractured, and only roughly grammatical, showing the continual changes of mind of the speaker as they monitor their speech, their perception of the impact of their speech upon the listener, and their own developing train of thought. Umming and erring and the use of phrases such as 'you know', 'I mean', and 'well, yeah' are placeholders that are necessary to keep the stage in a conversation while one's thoughts are put into order and the difficult task of converting them into speech in real time is executed.

Transformational grammar

Modern psycholinguistics is founded on work in the late 1950s and the 1960s of Noam Chomsky (see Topic A3). The behaviorist approach had treated language as a behavior like any other, subject to the same laws of learning and reinforcement. Chomsky argued that this could not account for the ability that children had to produce novel utterances, which they had never heard and had never been rewarded for producing. Later work showed that parents do not reward children for producing grammatically correct utterances, but do reward true utterances even if they are ungrammatical. Grammar, therefore, cannot be learnt through conditioning. Chomsky (1957, 1965) suggested that the **surface structure** of language (what is actually produced) was the end product of a series of **phrase structure rules** applied to a **deep structure**, which corresponded to the thought underlying the intended utterance. Most controversially, he argued that the deep structure was universal across all humans in all cultures, as were the classes of phrase structure rules; all that differed was the way that the rules operated to transform the thought into a surface form. A child does not need to learn the grammar of their language, because the structure of the rules is innate; what they do need to learn is the particular vocabulary of their language and the operation of the rules.

According to Chomsky, all sentences can be broken down into a phrase structure, such that a sentence consists of a **noun phrase** (NP) and a **verb phrase** (VP), shown in *Fig. 1*. The NP can be as simple as an article and a noun or a determiner and a noun (Art+N or Det+N, e.g. 'a/the mouse', 'this mouse'), or can include more complex adjectival constructions with secondary phrases (e.g. 'the nice furry mouse I had told you about earlier'). The VP includes at least a verb and another NP (e.g. 'ate' + 'the corn'), but can also be more complex (e.g. 'had rapidly eaten' + 'the corn'). The same deep structure can, by the

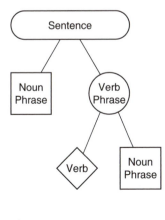

The Dog Chased   The Cat

Fig. 1. A simple phrase structure diagram. All sentences consist of a noun phrase and a verb phrase. The simplest verb phrase consists of a verb and another noun phrase, so the simplest sentences are two noun phrases linked by a verb.

application of different rules, lead to sentences with different surface structure but identical meaning, such as: 'the mouse ate the corn' and 'the corn was eaten by the mouse'. Although Chomsky was a linguist, and initially his theories were only meant to be applied to language syntax rather than to its production and comprehension, psycholinguists have searched for evidence about the psychological reality of deep structure and phrase structure rules.

Modularity

The psycholinguistic research that has followed Chomskian theory about language has become associated with the idea that cognitive functions are **modular**, that is, that each particular domain of cognitive ability has its own pattern of development, its own innate modes of operation, and its own local neural organization (see Topic A1). Broadly speaking a module is a processing system that works with specific classes of information from a specific domain, although it can receive information from earlier modules (e.g. perception) and provide information for later modules (e.g. those involved in thinking and reasoning). Because modules are assumed to be structurally innate, and to process information within themselves before passing it on, it makes sense for them to be localized within a discrete region of the brain.

Because the language that everyone learns depends upon the one that they hear around them as babies, and yet our domestic animals do not start talking to us or comprehending anything other than the simplest formulaic commands, both nature (genetics) and nurture (environment) must play a part in language acquisition and use. Chomsky had argued that human children have an innate 'language acquisition device' or LAD that gives them the specific ability to learn vocabulary from the speech they hear around them, and to derive the rules for forming phrase structures, for fitting lexical items into the structures, and for transforming the deep structures into surface structures. This seems to set language learning and use apart from other cognitive operations, and much debate focuses on the extent to which language use is dependent upon other cognitive capacities such as the span of the phonological loop in working memory (see Topic D2). The universality of deep structure and of the classes of phrase structure rules also implies an innate capacity, and hence a modular nature for language. Evidence from the development of language in children deprived of linguistic experience has been used in arguments about the degree to which language development is continuous or whether it passes through discrete stages; and whether there are optimal or critical periods during which exposure to language is necessary for neural development and hence cognitive development to support later language use.

Primate language

If human language abilities are supported by an innate capacity, then it would seem likely that species who are closely related to us genetically might also possess at least the rudiments of a language acquisition device. Several teams of researchers have spent many years providing young chimpanzees with a linguistic culture specifically designed to foster language development, with only limited success. Humans are able to produce a wide range of vocal sounds because of a peculiar physiological change that occurs in the neonatal throat, when the larynx descends to form the voice box. Although this allows us to produce voiced speech, it also means that, unlike chimpanzees, we cannot drink or eat and breathe simultaneously, so can choke to death on our food. Because this change does not occur in chimpanzees, they cannot produce anything like the range of human vocal sounds, and so are not able even to imitate human speech. In the wild chimpanzees do not use language, for their vocalizations are limited to the types of signaling used by many animals. The question for researchers has been to discover whether they possess the ability to learn elements of a language that has been specifically designed to make the most of their perceptual and productive abilities.

To provide a chimp with the right environment, researchers have to adopt a very young chimpanzee and surround it with a linguistic environment as it grows up. The chimp really has to become part of their family, with all of the inevitable subjectivity that this brings. The Premacks used a sign-board to teach a chimp called Sarah to manipulate symbolic shapes or tokens. Despite intensive training, Sarah did not spontaneously use the tokens to express herself, but would only use them passively, when a 'conversation' was started by a human (Premack, 1971). Other attempts by the Gardners (1969) with Washoe, and by Terrace and colleagues (1979), with Nim Chimpsky, used a hand- and finger-based gestural sign language called American Sign Language (ASL) that had been developed for deaf humans. It was hoped that this would be more suited to chimpanzees' productive capacity because they have excellent manipulative skills. After four years of training, Washoe had only learnt 132 signs, although she was able to teach these signs to another chimpanzee, Loulis, who learnt to use 50 signs without being taught them by a human.

Although the chimps in these studies learnt a signing vocabulary, and would produce utterances made up of strings of signs, the utterances did not show any sign of structure such as word order (i.e. a consistent order of subject, verb and object, or of noun and adjective). There is doubt as to the accuracy of their signing, and the degree to which it depended upon the chimp following the lead of the human researcher that they were conversing with, and upon the human researcher's subjective interpretation of a chimp's gestures. Pinker (1994) reports a deaf researcher skilled in ASL as being highly critical of the extent to which the chimps were actually able to sign. In double blind conditions, where the researcher signing to Washoe could not see her, and where the person transcribing Washoe's signing could not see the first researcher, her performance declined substantially. In a test of Nim Chimpsky's ability to take turns in a conversation, it was found that 71% of the time, the chimp signed at the same time as the human researcher.

Even under these optimal conditions, the chimpanzees' best performance is very poor compared with human children, who begin to form two-word utterances within a year of learning to talk, and who are freely producing complex and grammatically regular utterances a year after that. Savage-Rumbaugh has looked at language comprehension rather than production, in Bonobos, and has reported substantial comprehension abilities, even for complex phrases in which nouns are qualified by adjectives, and words like 'it' are used to refer to objects that have already been mentioned (Savage-Rumbaugh et al., 1988). The implications for the evolution of language in humans are that the capacity for language production developed after the abilities for language comprehension, which must therefore have served some other purpose (such as inferring causality of events or the existence of hidden events from sounds in the natural world). Grammatical language production seems unique to humans, and its development may have driven the evolution of physiological changes in the mouth and throat, and perhaps even changes in the structure of the brain to support other cognitive abilities.

H2 DISCOURSE

Keynotes

Pragmatics	Much language is ambiguous, even though entirely grammatical, and depends upon the context to make sense. Providing a suitable context can improve the comprehension and recall of information, because it can be related to a memory schema to add implicit details. Speakers relate new information to previously given information (the given–new strategy). This allows new information to be associated with the currently active schema.
Co-operation	Conversations are co-operative exchanges in which speaker and listener attempt to exchange information efficiently. Grice posed the four co-operative maxims of quantity, quality, relevance and manner that a speaker should follow to allow a listener to assume that they were being appropriately informative. Clark's idea of common ground is that speakers attempt to establish shared knowledge early in a conversation.
Turn taking	To avoid talking at the same time as each other, and to alert people to the start and end of conversations, people use conventional phrases to open and close a conversation and follow conventions in deciding whose turn it is to speak next. Breaking these rules makes a speaker sound rude or inconsiderate.

Related topics	Communication (H1)	Language development (H3)

Pragmatics

A feature of human language use that has bedeviled attempts to develop artificial systems for language comprehension such as automatic translators and speech-to-text devices, is the tendency for the same utterance to mean radically different things depending upon the context or the **pragmatics** of the situation. 'Give me the key' means different things if said by a policeman to a drunk driver ('pass me the device that allows you to operate the car, which you are not allowed to do in your current state') or by a singer to a conductor ('ask the orchestra to play a note so that I know how to start singing the song and then I will start'). Taken out of context, language can be perfectly grammatical and meaningful on a phrase-by-phrase basis, but still mean nothing:

> 'The procedure is quite simple. First you arrange items into different groups. Of course one pile may be sufficient depending upon how much there is to do. If you have to go somewhere else due to lack of facilities that is the next step; otherwise you are pretty well set. It is important not to overdo things. That is, it is better to do too few things at once than too many. In the short run this may not seem important but complications can easily arise... After the procedure is completed one arranges the materials into their appropriate places. Eventually they will be used once more and the whole cycle will then have to be repeated. However, that is part of life.'

This is an extract from a passage devised by Bransford and Johnson (1972). Although it is grammatical, and each sentence makes some sense, people who read the entire version thought that it was incomprehensible and could only recall an average of 2.8 chunks of information, whereas those who had been told beforehand that it was about washing clothes thought it made perfect sense, and could recall on average 5.8 chunks. This study showed that having the appropriate memory schema activated contributed to the comprehension and retention of information (see Topic D3). The memory schema allows listeners (or readers) to add in information that had not been explicitly given in the discourse (such as that the 'items' were clothes, for example, or that the 'facilities' were washers and dryers).

Similar work has shown that material that fits with expectations is actually comprehended faster than material that requires the listener to work out the relationship to what has been said before. Haviland and Clark (1974) proposed that speech is often organized on a **given–new** basis, with a speaker making explicit effort to relate novel information to something that they know the listener has just been told. For example, it is easier to understand and recall the sentence 'It is my birthday tomorrow, so I must buy some balloons, candles and cake' than 'I must buy some balloons, candles and cake, because it is my birthday tomorrow'. The former sentence sets up the situation first, allowing the party schema to be activated, and then mentions items that fit the schema. The listener can identify the given information rapidly, and use this to resolve any ambiguity about the new information. The second sentence requires the listener to hold the shopping list in working memory until the party schema is activated, increasing the chance that they will forget or misinterpret something.

Co-operation

Grice (1975) pointed out that conversations are co-operative exchanges in which the speaker and the listener both make an effort to be meaningful. The speaker attempts to provide meaningful information, and the listener refrains from willfully misunderstanding them (i.e. when the speaker is ambiguous, they do their best to infer the most likely or polite meaning). He proposed four maxims that a speaker should heed:

- Quantity: provide as much information as is needed, but not more.
- Quality: provide information that is true.
- Relation: provide information that is relevant to the situation.
- Manner: provide information in a manner that is easy to follow.

If the listener can assume that these four **Gricean maxims** are being followed, then understanding what the speaker is saying can be simplified, because they know that everything in the utterance is important, true and necessary for them to know. To follow them, though, the speaker and the listener need to know to some extent what the other one already knows: if they did not, then the speaker might include irrelevant details to set the scene for the listener, or details about the topic that the listener already knows. Clark (1996) discusses the idea of **common ground** in co-operation generally, and particularly in conversation. This consists of all of the shared beliefs and knowledge that people in the domain in which the conversation is being conducted can be assumed to have. Common ground must be established early in the conversation by specifying the general topic or domain of the discussion (e.g. 'You know the problems we were having yesterday, well…'), and if either speaker feels that something is going adrift in the conversation, then they make explicit attempts to re-establish common ground (e.g. 'wait a minute, you mean...').

Turn taking

In normal conversation, people do not talk at the same time, but co-ordinate their speech in a **turn taking** series. A speaker normally continues speaking until they have had their chance to reach the end of their utterance, filling any pauses due to uncertainty about phrasing or vocabulary with non-words such as 'umm' or 'err', or phrases such as 'like', or 'you know'. These fillers vary according to the language being spoken, and the register (the style of speaking). In some languages (including English) intonation can be used to signify an end point; many cultures also include facial and hand gestures to show that the speaker has finished their turn. Sacks, Schegloff and Jefferson (1974) proposed three conversational rules that speakers followed when deciding who should speak next. First, the person who had been directly addressed by the previous speaker should speak next. If this rule did not apply, then anyone else other than the speaker could speak, and whoever spoke first should continue speaking if someone else also tried to speak. Finally, if no one else started speaking, the first speaker could start another turn.

Sacks, Schegloff and Johnson also addressed the conventions that people used to start and stop conversations. We do not simply start right in on a topic, nor walk away when our meaning has been conveyed, but use social conventions to signal to other people that we want to have a conversation, and also that we have finished talking and want to do something else. Opening phrases take the form of a summons–answer sequence (e.g. 'excuse me'/'what?') where the first person must wait for the second to respond before continuing, or be thought rude, and the second must respond. If they cannot get involved in a conversation they have to indicate so somehow, and not just ignore the first person. In using the telephone, the ringing takes the form of the summons, and the person being called has the first chance to speak, but in a conventional manner such as giving their name or just saying 'Yes?' The second response is usually conventional, so if the first takes the form of an enquiry (e.g. 'How are you?') it is obtuse for the second person to give a factual answer (e.g. 'I have a pain in all the diodes down my left side') instead of the conventional (e.g. 'Fine, you?'). Ending a conversation is also ruled by conventions, but is more complex because there are two stages: agreeing to close the conversation, and then actually closing it. When one person uses a phrase that signals their desire to end the conversation (e.g. 'Well, okay then...') the other is obliged to either agree by mirroring their phrase (e.g. 'Okay...') or by introducing a new topic (e.g. 'So, did you hear about...'). Although the exact phrases differ from language to language, the patterns are universal.

H3 LANGUAGE DEVELOPMENT

Keynotes

First words	Newborn children make a wide range of vocal sounds, eventually restricting them to the sounds that they hear in the language around them. Their utterances become more repetitive and word-like. By their second year infants are systematically using single words or holophrases. The roles that the words represent follow a developmental sequence: agents, actions, objects, states, associates, possessors and locations. By 2 years old, most children are using patterned speech including two-word phrases.
Learning grammar	The speed of language learning varies, but the order of acquisition is fairly constant. Children begin with words and grammatical forms that describe things in the here and now, then add past irregular forms, and then past regular forms. At this point, they often over-generalize and incorrectly attach regular endings to irregular verbs that they had previously used correctly, such as 'wented'. They can also generate the correct forms for words they have never heard before such as 'one wug, two wugs'. The order in which grammatical forms are acquired depends upon the complexity of the ideas represented, and the complexity of the speech needed.
Deprivation	Deaf children readily develop language in a gestural form. If raised by non-signing parents, they spontaneously develop their own grammatically structured gestures; if raised by signing adults they learn the signing system used by their parents, even babbling manually. Receiving linguistic input is more important than the modality of that input. Blind children, who cannot perceive distant objects being talked about, have initial difficulties acquiring particular grammatical forms such as pronouns and referents, but later catch up. The infant mind is prepared to learn language, but input is needed.
Critical period	Children raised in isolation without language are rare, and may have other deficits in addition to the absence of language, but the limited evidence suggests that if they are exposed to a linguistic environment before puberty they can catch up and develop normal language skills. After puberty, their abilities decline, and they will be less able to speak grammatically and to form complex sentences. A similar constraint applies to learning a second language, with young children reaching a level of grammatical fluency equal to single language speakers, while older children make more grammatical errors. Evidence suggests there may be an optimal period beyond which ability declines, rather than a critical period.

Related topics	Discourse (H2)	Cognitive development (J1)
	Speech (H4)	Nature and nurture (J3)
	Language and thought (I3)	

First words

Infants communicate with their parents from the moment that they are born. At first they cry, but within a few weeks they are cooing and then babbling, as if they are exploring the range of sounds that they can make. At first they generate all sorts of sounds, including those that are not part of their parents' language, but over the next months, their babbling seems to home in on the sounds that their parents recognize, becoming more word-like and repetitive, including sequences in which they produce the same sound over and over, 'gagagaga' rather than 'baoapagu'. Escalona (1973) identified a sequence of interactions in which parents give objects to their very young babies, naming them and talking about them. By 6 months, the babies are picking the objects up and showing them to the parents, making vocal noises as they do so. At 8 months, parents are pointing at objects to direct their babies' attention to them, and within a few weeks the infants start pointing to things when someone else is there.

By the time that they are a year old, infants are beginning to associate single word utterances or **holophrases** with general classes of event, and to use the same sound systematically. Greenfield and Smith (1976) identified seven aspects or roles that these early one-word utterances related to, and argued that their use followed a developmental sequence. First, infants name the **agents** who are doing things – usually people. Then they begin to name the **actions** that the agents are carrying out, or the state that results from these actions. Then they start to name the **objects** affected by the actions, followed by the **state** of those objects after the action. They then in turn start naming the objects that are **associates** of the action, **possessors** of objects, and **locations** of objects. Each of these utterances has a distinct semantic status, and they can be seen as building blocks for a grammatical sentence. At around 18 months to 2 years, patterned speech begins, with pairs of single words being produced closely together to describe a single situation. This is often called the 'two-word utterance' stage, because parents notice that their child is now producing pairs of words that are recognizably relevant to the situation, but their speech still includes many single words, and indeed also longer chains of three or four words.

Learning grammar

The rate at which children learn to talk varies widely, but all children who are brought up in normal surroundings and who do not have some specific disorder do learn to speak. What does seem to be consistent is not the speed at which they learn, but the order of acquisition of what they can say. To examine the way that children progressed from using single words to using these words grammatically, Brown (1973) studied the language development of three English-speaking children; Adam, Eve and Sarah. He identified 14 morphemes, each corresponding to a grammatical unit, case of a noun, person or tense of a verb. Although the three children varied in the ages at which they learnt to use these morphemes, they did so in a fairly consistent order that was not related to the frequency with which their parents used them (*Table 1*).

A noticeable feature of this order of acquisition is that the past irregular form of common verbs is learnt before the past regular form, so a child will learn to say 'I went' before 'I turned' or 'I spilled'. Once the regular form of the past tense has been learnt, though, children often overgeneralize it and begin to make mistakes by applying it to irregular verbs, saying things like 'I goed', or even 'I wented'. This has been taken as evidence of rule learning: first the child learns a few one-off irregular forms that they hear a lot, that introduce them to

Table 1. The order of acquisition of fourteen morphemes of English, identified by Brown from a study of four children

Morpheme	Example
1. Present progressive	*verb*-**ing**
2. Preposition of containment	**in**
3. Preposition of superposition	**on**
4. Plural	*noun*-**s**
5. Past irregular	e.g. **went**
6. Possessive	-*noun*-**'s**
7. Uncontractible copula of be	These **are** *adjective,* This **is** a *noun*
8. Articles	**a, the**
9. Past regular	*verb*-**ed**
10. 3rd person regular	*verb*-**s**
11. 3rd person irregular	**has, does**
12. Uncontractible auxiliary of be	He **was** *verb-ing,* **Were** *noun-s verb-ing?*
13. Contractible copula of be	They-**'re** *adjective,* That-**'s** a *noun*
14. Contractible auxiliary of be	They-**'re** *verb-ing,* I-**'m** *verb-ing*

The morphemes are shown in bold face in the examples, with the grammatical classes of other words in italics.

the idea of a past tense; they then learn the past tense forms of some less common verbs that share the –ed ending, and infer that all verbs can be made to relate to the past by adding this ending; and then have to learn that some verbs are exceptions. Berko (1958) tested children's ability to apply grammatical rules to novel words by teaching them nonsense words for imaginary animals such as **a wug**, or actions such as **ricking**. When shown pictures of situation children would easily generate answers including the plural form **two wugs**, the possessive form **a wug's hat**, the past form **he ricked** and the present form **he ricks**.

It is important to remember that these rules account only for the child's production of language; even 1-year-old children can understand a lot more than they can produce. Golinkoff *et al.* (1987) showed that infants who were only producing single words could distinguish between sentences such as 'A is **gorping B**' and '**A is gorping with B**', where the non-word **gorp** corresponded to the action of turning. Two features seem to govern the sequence with which aspects of language are learnt: the cognitive complexity of the ideas that are being expressed, and the formal complexity of the rules needed to form the surface structure of the expression. Cognitively simple ideas such as position and number come before more complex ideas about time and possession, and formally simple expressions such as adding a whole word (in, on) or an ending (–'s) come before more complex expressions such as the tenses of verbs. Simple rules and concepts provide the foundation for more complex rules to be learnt later. Slobin (1973) compared 40 different languages, and suggested six principles that children seemed to use in the patterns of acquisition of grammatical forms (*Table 2*). The first three of these relate to the complexity of the ideas or meanings being expressed, the second three to the complexity of the surface structure of the expressions.

Deprivation

Children who are born deaf can see their parents' lips move, and can see the gestures that they make to objects, but they cannot hear the sounds of language. The way that these children learn to interact with their parents provides

Table 2.   Six principles used by young children in learning a language

1. Identify systematic modifications in word forms
2. Identify grammatical forms that relate to semantic distinctions
3. Avoid exceptions to rules
4. Attend to the ends of words
5. Attend to the order of words
6. Avoid interrupting or rearranging phrases

compelling evidence for the innateness of linguistic ability. Pettito and Marentette (1991) compared two congenitally deaf infants born to deaf parents who used American Sign Language (ASL) with three hearing babies born to hearing parents who spoke normally. The deaf infants' hand movements included 32% and 71% of manual 'babbling', compared to a maximum of 15% in the hearing babies, and almost all of the deaf children's manual babbling could be identified as units of ASL. Goldin-Meadow and Mylander (1998) compared the spontaneous gesture systems developed by four deaf US and four deaf Chinese children, all with hearing mothers who did not use sign language. They found that the children formed gestural sentences that were more complex than those made by their mothers, and that the gestural systems used by the US children were less like those used by their own mothers, and more like those used by the Chinese children. In Nicaragua, deaf children who had been raised by hearing parents before attending a school for deaf children where lip reading was taught persisted in using their 'home grown' signing, and developed their own form of sign language, which they taught to new children joining the school (Senghas and Coppola, 2001).

These studies show that language development is not tied specifically to the vocal modality, but that the abstract structure of language can be acquired in other modalities. Furthermore, the structure of the language seems to follow the same patterns as spoken language. The experience of perceiving a structured language is more important than the modality of that language. Blind children who can hear language, but who cannot see the objects at a distance that are being referred to, do learn to talk, but initially have difficulty understanding pronouns and referents such as 'this' and 'that' (Landau and Gleitman, 1985). The evidence supports the idea that there are processes ready in the infant mind for language learning, but that linguistic input is needed to give these processes material to operate upon.

Critical period   Children who are not blind or deaf, but who are raised in isolation without any linguistic input provide another source of evidence for the interaction of innate and environmental influences on language development, and in particular for the argument that there is a critical period in which language must be learnt. The evidence is limited and controversial, for the circumstances in which these children find themselves are bizarre and can never be compared to a controlled experiment in which only one or two variables are manipulated. In India, there are cases in which children abandoned by their parents appear to have been raised by wolves: in the 1920s, Kamala and Amala; and in the 1970s, Ramu. None of these children learned to speak more than single word utterances. There might have been reasons why these children had been abandoned, however: they might have been retarded, or had retarded parents, and so had limited linguistic ability for other reasons.

In the US, a girl called Genie was discovered at the age of about 14, having been kept tied in a chair without being spoken to since before she was 2 by a father who treated her like a dog. She was unable to speak. Once taken into care, and given a richer linguistic environment, she did learn to use words, and to form simple sentences with regular word order – but her speech was far from normal, since it lacked functional words such as **a**, **the**, **and**, and **but**, and she was unable to combine her simple phrases into more complex sentences (Curtiss, 1977). A younger child known as 'Isabelle' was born to a deaf mother who did not look after her, and was raised in isolation until she was 6. Within a year of her discovery, she had caught up with normal 7-year-olds, and was able to attend a conventional school (Davis, 1947). These two cases suggest that the age at which linguistic input is made available to a child is important: if the child has reached puberty, then their ability to learn language diminishes rapidly. Before puberty, a child remains able to acquire language even if initially deprived of linguistic input.

This critical period is also evident in second language learning. Adults who learn a second language can rapidly acquire vocabulary and can learn to apply the different grammatical rules to speak and understand the second language, but often retain an accent, and when taken out of the environment of the second language, start to forget it. Children who have started to speak one language and who are then confronted with a second language have a much slower rate of progress, and often appear to be impaired in their use of their first language. Within a year or two, however, they speak both languages fluently. The critical age seems to be around seven: in one study, Korean and Chinese immigrants to the US were tested on English grammar (Johnson and Newport, 1989). Those who had arrived in the US before they were seven performed identically to children who had been born in the US, but the older that children were when they arrived in the US, the less well they were able to learn English grammar. All of them were acceptable speakers of English, however. A similar pattern of results has been found in deaf children who are only introduced to ASL some time after birth (Newport, 1990), and like Genie, the errors that late learners of ASL make include the omission of function words. Again, the fact that they do learn to use ASL suggests that the critical period is not an all-or-nothing window beyond which language cannot be acquired, but that it is more like an optimal period, beyond which the ability to acquire new linguistic forms declines.

H4 SPEECH

Keynotes

Continuous stream	Speech is a continuous stream of sound containing vowels and consonants (phonemes), which form morphemes (units of meaning). These chunk into words and phrases. We can recognize words in our own language very rapidly, but an unknown language sounds unstructured. This indicates that speech perception has a top-down component in which we anticipate the words that a speaker will use and search for them in the sound stream.
Categorical perception	Vowel sounds differ on a continuum, yet are perceived as discrete and different sounds. Consonants are only easily recognizable when combined with a vowel, and yet this changes the way that the consonant sounds. Vowels and consonants are subject to categorical perception, and closely related stimuli either side of a categorical boundary can be discriminated, while widely different stimuli that are within the same category cannot be discriminated. Once categorical boundaries have been learnt, it is difficult to discriminate sounds in languages that use different categories.
Multimodal speech	People deprived of auditory input can learn to perceive speech by lip reading. Hearing people can also be influenced by lip movements, as in the McGurk effect, where the sounds of one phoneme are dubbed onto the lip movements for another. What is heard is neither original phoneme, but a blend of the two. The ventriloquist effect occurs when we attribute sounds from one source to another speaker's lips, even if they are spatially separated.
Speech slips	The nature of the errors people make in speech can tell us about the levels of processing involved in speech production. Substitution of semantically related words indicates that there is a functional level; swapping of words within a sentence shows that there is a positional level; using the wrong word endings shows that there is a phonetic level; and spoonerisms show that there is an articulation level. In tip-of-the-tongue experiences we are often correct about phonological aspects of the target word, even though we cannot grasp all of it, suggesting that different aspects of articulation are supported by separate processes or stores.
Aphasias	Brain injuries on the left side of the head often affect speech. Injuries to Broca's area result in halting speech, with problems at the phonemic and articulatory levels. Damage to Wernicke's area allows fluent but meaningless speech, with problems at the positional and functional levels. Damage to the connections between the two areas prevents people repeating what has been said to them. Damage to links between either of these two areas and the rest of the brain leads to the usual pattern of problems for the area, but allows people to repeat speech that they hear.

Related topics	Auditory perception (B2)	Concepts (E3)

Continuous stream

Speech is a continuous stream of sound. Although when we speak we feel that we are saying individual words, and that each word is pronounced in a particular way, analysis of the acoustic energy patterns that we make shows that words are run together (**co-articulation**), and the way that a letter sound is made depends on its place in a word, the sounds coming before and after it, and even the place that its word has in a phrase. These problems of **linearity** (the effect adjacent sounds have on each other), **non-invariance** (different sounds lead to the same perception) and **segmentation** (how we perceive the stream into units) are challenging for models of speech perception.

The speech stream can be thought of as a **hierarchical structure**, consisting of vowel sounds and consonants, called **phonemes**, which are chunked into **morphemes** (e.g. '–ing' or 'pre–'), segments that have a distinct semantic or grammatical status. These morphemes cluster into words, which form phrases, which form sentences. When you listen to your own language it is hard to believe that it does not consist of discrete words separated by gaps, but if you hear a foreign language, that you cannot speak, you will find it impossible to identify individual words, although there may be pauses at the ends of phrases or sentences.

Despite this, we are remarkably good at perceiving words in the speech of people who are speaking our own language. In fact, we seem to be able to identify the words being spoken before they have actually been completed. Marslen-Wilson and Tyler (1980) asked people to listen to a passage of speech and to press a button whenever target words occurred: the average duration of the words was 370 msec, but the button presses came on average only 275 msec after the start of the target words. This sort of evidence suggests that we are not perceiving speech by a 'bottom-up' process alone, in which we perceive individual phonemes and build them up into morphemes, and then words, and so on. To recognize a word before it has been completed, there must be some additional 'top-down' process which is anticipating the phrases and words that are about to be said, and searching the speech stream for sounds that fit candidate words.

The **phonemic restoration** effect gives evidence of the action of top-down processes in speech perception. Warren and Warren (1970) played volunteers tapes from which they had replaced a particular phoneme with a hiss of white noise. When the word containing the missing phoneme was embedded in a sentence, their volunteers reported hearing a complete word, but the word that they heard depended on the sentence (*Fig. 1*).

Categorical perception

Vowel sounds are made by vibrating the vocal cords and changing the position of the tongue and the back of the throat to vary the way that the sound resonates in the nasal cavities. Because all the muscles involved can move smoothly from position to position, there are no clear cut definitions of what a particular vowel sound is, and vowels can be combined into diphthongs, in which the sound pattern flows smoothly from one vowel to another rather than changing discretely. For example, the word 'eye' is a diphthong produced by a short /a/ sound followed by a long /i/ sound (putting a letter between slashes is a conventional way of printing the sound rather than the name of the letter).

Different languages may have different numbers of vowel sounds, and different regional accents within those languages will vary the modal position of each vowel. The duration of vowels can also vary and this can be important in some languages but not in others. Someone who learns a second language with

Fig. 1.   *In the phonemic restoration effect, the missing phoneme is reinstated by the listener, in accordance with the context in which it occurs, even though the context occurs after the missing phoneme.*

different vowel sounds to their own will find it very difficult to learn to produce these sounds, and will have a persistent accent. Imitating a foreign or regional accent is largely a matter of mimicking the characteristic vowel sounds that those people bring to your language.

Consonants are made by blocking the flow of air by parts of the mouth or throat, or by placing the tongue against the teeth, cheeks or roof of the mouth. Air can be stopped and then released, or allowed to squeeze through narrow channels. Because of the more specific nature of the mouth parts and explosive or fricative nature of consonants, they are more discrete than vowels, but still fall into families of confusable sounds. In addition, positionally identical consonants can be voiced (/b/, /d/, /g/) or unvoiced (/p/, /t/, /k/) depending whether they do use or do not use the vocal cords. A consonant sound edited out of a speech stream and presented alone does not sound like a consonant, but as a brief click or buzz. In general they are only recognizable in conjunction with a vowel sound, such as /di/ or /du/. In such combinations, the part of the sound corresponding to the /d/ varies according to the vowel that is to follow. The same perception, /d/, is thus made from different sounds.

A key attribute in distinguishing between voiced and unvoiced sounds such as /b/ and /p/, or /d/ and /t/, is the **voice onset time** (VOT), the delay between the start of the consonant sound and the start of the vocal cord activity associated with the following vowel. If it is less than about 40 msecs, the consonant is perceived as voiced. If it is longer than 60 msecs, it is perceived as unvoiced. In normal speech, the VOT varies widely, and can be anything from 10 msecs up to 100 msecs, but people cannot distinguish between two sounds where the VOTs are both shorter than 40 msecs, or are both longer than 60 msecs. Both sounds will seem to be identical. If the VOTs fall either side of this boundary, then the distinction is readily made. One will sound voiced, the other unvoiced (Lisker and Abramson, 1964).

Evidence like this, and the way that continually varying vowel sounds are pigeonholed into a small number of discrete sounds, suggests that phonemes in speech are identified by a form of **categorical perception**. This allows a range of varying stimuli to be perceived as a small number of categories, but once a stimulus has been categorized, it is impossible to distinguish it from other members of the category, even if it is from one end of the boundary and the others are from the opposite end. Categorical perception also occurs in color perception, where light energy can vary continually from infra-red to ultra-violet, and we can see everything in between, but prefer to identify bands of the spectrum as in a rainbow. A green that is almost yellow is still green, and a green that is almost blue is still green; both seem more similar to each other than to the yellow and blue that they are actually physically more similar to. The same effect happens with sounds, and allows a speaker to vary attributes of the sounds that they are making to minimize the mouth movements needed to speak, while still allowing a listener to identify discrete phonemes. Once the categorical boundaries of one's language have been learnt, it is difficult to learn to discriminate the sounds of languages that use different categories (Werker, 1995): English speakers, for example, have difficulty discriminating the several different /sh/ and /ch/ sounds used in Chinese.

Multimodal speech

Infants who are learning their first language can often be seen paying close attention to their parents' lips, and people can learn to lip read. To some extent, we are all influenced by what we see another person's lips doing when they are speaking, as is shown by the **McGurk effect** (McGurk and McDonald, 1976). In this illusion, the sound of someone saying one consonant-vowel combination (e.g. /ga/) is dubbed onto the lip movements of someone saying a different consonant vowel combination (e.g. /ba/). With your eyes closed, it is easy to hear that they are saying /ga/, but as soon as you open your eyes, the sound changes: not to /ba/, but to /da/ – neither one sound nor the other, but a blend. Speech perception is multimodal, taking input from both the visual and auditory senses.

One explanation for how this multimodal perception occurs has been proposed by Massaro, in his Fuzzy Logical Model of Perception (Massaro, 1998). There are three stages in this model: the evaluation of the quality of each sensory input; their integration into a single representation; and then a decision about what sound this represents. Because we are having to make perceptual decisions in a noisy world, and one or other source might be more or less available at different times, this allows the perceptual decision making to weight the individual sources differently according to their quality. In an experiment, five sounds varying from /ba/ to /da/ were presented in all combinations with five lip movements for the same sounds. When both sources corresponded exactly to /ba/ or to /da/, identification was nearly perfect. When one or either was intermediate, however, the probability of identifying the sounds as one thing or the other changed: not suddenly as one might expect with a pure categorical perception, but smoothly, showing that listeners would sometimes say one thing and sometimes another for the same combination. Although the individual sensations might be perceived according to a categorical process, their combination was not so all or nothing.

Another source of evidence for the blending of visual and auditory inputs in speech perception is the **ventriloquist effect**, whereby speech coming from one side of a speaker's face is perceived as coming from the area of their mouth. In cinemas and at home when watching television, we are seldom aware that the sound is coming from a few speakers located around the room, rather than from the faces we see on the screen. Driver (1996) showed that placing a black rectangle over the lips of a person speaking some words could abolish the ventriloquism effect, making listeners aware of the true origin of the sound.

Speech slips

When we speak, we have to convert our thoughts into vocal output, through a number of stages. The occasional mistakes that people make in producing speech have been used as evidence for the identity of the stages, for there is a systematic pattern to the types of errors or slips that are made. Garrett (1980) identified four general classes of speech error (*Table 1*), each one indicating an error at a different stage of speech production. Garrett's stages comprise a **functional level**, at which the concepts are instantiated as lexical items and grammatical roles; a **positional level**, at which syntactic structure of the sentence is organized, so word order is determined; a **phonetic level**, at which the lexical identity of each word is inserted into the sentence; and an **articulation level**, at which the motor commands are generated to produce the utterance.

Table 1.   Four classes of speech slip

Slip	Description
Word substitution **Functional level**	Using a wrong but semantically related word (e.g. cat for dog; tall for short)
Word exchange **Positional level**	Exchanging units from the same grammatical category within a sentence with each other (e.g. subject and object: 'Hit the **hammer** with the **nail**')
Morpheme stranding **Phonetic level**	Positioning verbs or nouns with the wrong grammatical units, which remain in their correct position (e.g. '**sixt**ing **swing**ies', '**Nail**ing the **hammer**')
Sound exchange **Articulation level**	Spoonerisms, confusing sounds from nearby words, either exchanging whole phonemes or repeating features from one phoneme in another (e.g. '**h**issing your **m**ystery lecture', '**L**ead **l**orry, yellow lorry')

Sometimes, we cannot immediately find the right word to describe something, and experience an annoying tip-of-the-tongue feeling. This has been attributed to a failure at the Phonetic level, since we know what we mean but just cannot bring the correct word to mind. Brown and McNeill (1966) gave people definitions of rare words and when they could not immediately answer with the correct word, asked them to say what letter it started with, how many syllables it had, other words it sounded like, and for other words with similar meanings. They found that people got the first letter correct for 62% of the words, and the correct number of syllables for 57%. Similar words that were generated usually shared the stress pattern and several phonemes. This shows that people were getting some access to an incomplete phonological representation, and that perhaps different aspects of the word's articulation, such as its stress pattern or prosody, and phonemic constituents, are generated or stored by different processes. Errors known as **malapropisms** follow from errors at this stage of speech production, leading to a word being replaced by one that is similar but (often comically) incorrect. In an example from literature, in Act 4 Scene 2 of Shakespeare's *Much Ado About Nothing*, Dogberry says 'O Villain! Thou wilt be condemned into everlasting **redemption** for this. Come, let them be **opinioned**.'

Aphasias

Problems with speech production are known as **aphasias**, and are usually related to a brain injury to the left side of the head, just above and forward of the ear, supporting a general argument for localization of function, one of the predictions of modularity (see Topic A1). There also seems to be a systematic relationship between the type of speech problem and the specific area of the brain that has been damaged, providing further confirmation for the idea that there are different processes involved in speech production. Damage to **Broca's area**, a small region to the rear of the left frontal lobe, results in difficulty in speech planning and production at the phonetic and articulation levels. Speech is sparse and halting, with grammatical morphemes and function words omitted, although understanding is not badly affected. In contrast, damage to **Wernicke's area**, in the rear of the left temporal cortex, results in speech that is fluent but contains many semantic errors, and is often meaningless, suggesting damage to the functional and positional levels of production. Understanding speech is severely affected, suggesting that these levels are also involved in comprehension of speech. Damage to the tissues connecting the two areas, the arcuate fasciculus, leads to a syndrome called **conduction aphasia**, in which people cannot correctly repeat what has been said to them, although they can speak and understand speech normally. Damage to the tissues connecting the general speech area of the left hemisphere with the rest of the frontal and parietal lobes leads to **transcortical aphasias**. If connections to Broca's area are affected, people cannot generate their own fluent speech but can understand and fluently repeat what has been said to them; if Wernicke's area is disconnected, then a person can generate incoherent speech and can also repeat correctly what has been said to them, but cannot understand the words. The evidence from lesion studies is not unanimous however, for many people do not fit these clean-cut distinctions, and imaging studies on unaffected people do not always show the patterns of blood flow that would be expected for tasks requiring the functions associated with the areas. One possibility is that the areas provide working memory capacity for the functions rather than carrying out the functions directly.

A common problem in speech production that is not associated with a brain injury is stuttering, where a speaker apparently cannot produce a particular phoneme or class of phonemes without great effort. Some stutterers find that they can overcome their problem when in formal situations, in which they are able to speak in a controlled and planned manner; others find that anxiety makes matters a great deal worse. An age-old 'cure' involves speaking with a small stone in the mouth; the effort required to manipulate the stone seems to divert attention away from monitoring one's speech. Altogether, this suggests that the difficulty is not so much in producing speech as in overcorrection of speech. Lee (1950) was able to induce stuttering in fluent speakers by playing their own voice back to them over headphones, with a very small delay. If the delay was shorter than 50 msec, there was no problem, but at longer delays people began to pause and stutter, as if they were detecting and attempting to correct incorrect voice-onset times or other co-coordinated aspects of pronunciation. The problem was worst at 200 msec delay. As the delay approached 300 msec, the problems faded away again, as if the heard speech was now so different to the produced speech that it no longer interfered. Artificially generated non-speech sounds had a similar effect, if the sounds were structurally related to the speech and then delayed (Howell and Powell, 1987).

11 UNDERSTANDING WORDS

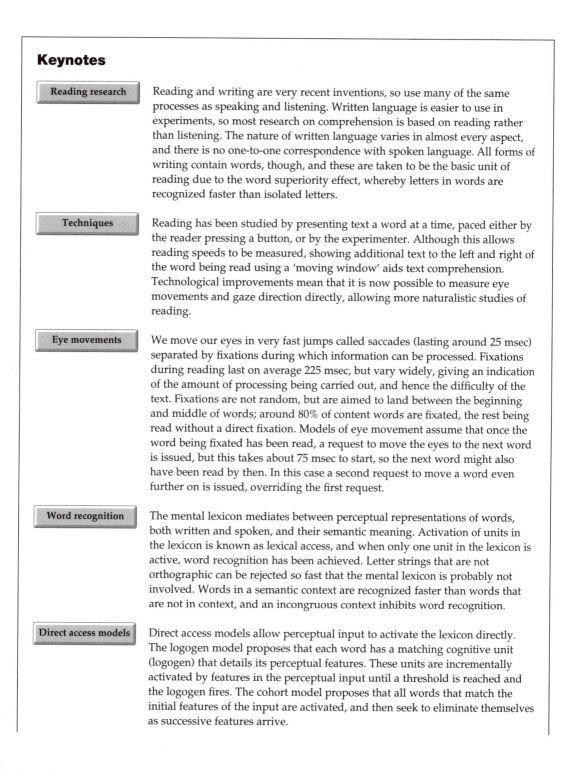

Keynotes

Reading research

Reading and writing are very recent inventions, so use many of the same processes as speaking and listening. Written language is easier to use in experiments, so most research on comprehension is based on reading rather than listening. The nature of written language varies in almost every aspect, and there is no one-to-one correspondence with spoken language. All forms of writing contain words, though, and these are taken to be the basic unit of reading due to the word superiority effect, whereby letters in words are recognized faster than isolated letters.

Techniques

Reading has been studied by presenting text a word at a time, paced either by the reader pressing a button, or by the experimenter. Although this allows reading speeds to be measured, showing additional text to the left and right of the word being read using a 'moving window' aids text comprehension. Technological improvements mean that it is now possible to measure eye movements and gaze direction directly, allowing more naturalistic studies of reading.

Eye movements

We move our eyes in very fast jumps called saccades (lasting around 25 msec) separated by fixations during which information can be processed. Fixations during reading last on average 225 msec, but vary widely, giving an indication of the amount of processing being carried out, and hence the difficulty of the text. Fixations are not random, but are aimed to land between the beginning and middle of words; around 80% of content words are fixated, the rest being read without a direct fixation. Models of eye movement assume that once the word being fixated has been read, a request to move the eyes to the next word is issued, but this takes about 75 msec to start, so the next word might also have been read by then. In this case a second request to move a word even further on is issued, overriding the first request.

Word recognition

The mental lexicon mediates between perceptual representations of words, both written and spoken, and their semantic meaning. Activation of units in the lexicon is known as lexical access, and when only one unit in the lexicon is active, word recognition has been achieved. Letter strings that are not orthographic can be rejected so fast that the mental lexicon is probably not involved. Words in a semantic context are recognized faster than words that are not in context, and an incongruous context inhibits word recognition.

Direct access models

Direct access models allow perceptual input to activate the lexicon directly. The logogen model proposes that each word has a matching cognitive unit (logogen) that details its perceptual features. These units are incrementally activated by features in the perceptual input until a threshold is reached and the logogen fires. The cohort model proposes that all words that match the initial features of the input are activated, and then seek to eliminate themselves as successive features arrive.

Serial access models	Serial models propose that a subset of words is activated on the basis of initial perceptual input, but then that each word is compared in turn against the complete input, and the first match is chosen as the recognized word. In Forster's search model, words are compared in order of frequency, so that common words are found sooner than rare words.
Dual access models	The phonological recoding hypothesis proposes that letter patterns are converted to sound patterns, and that reading then proceeds as if the input had been heard. Kleiman's model includes two routes, with phonological recoding being important while learning to read, and visual features taking over as skill develops. Liberman's motor theory argues that we understand speech by mentally preparing the articulatory actions that we would have to make to produce the sounds that we are hearing.
Related topics	Auditory perception (B2) Language development (H3) Communication (H1) Understanding sentences (I2)

Reading research

Reading is a very modern skill, in terms of human evolution, because writing is thought to have been invented no more than 6000 years ago. Until very recently, most people were illiterate, and had no need to read or write. The different modes of presentation of spoken and written language mean that there must be some initial differences in processing the sensory data, but it is reasonable to assume that language comprehension skills are based on hearing spoken language, and that reading uses many of the same processes. Because it is easier to control stimulus materials and their presentation in written language, most research on language comprehension has studied reading rather than listening.

There are of course important differences in the spoken and written forms of language. Whereas speech presents a continuous stream, which the hearer must segment, with some indication of phrase boundaries, most writing systems present discrete units at some level. The level varies widely between different writing systems, with letters corresponding to phonemes, symbols corresponding to syllables or morphemes, and ideographs representing entire words. Printed alphabetic scripts represent each symbol individually, whereas cursive scripts join symbols together, but both forms usually include word boundaries and punctuation to mark phrase boundaries. Some scripts such as Chinese and Japanese do not mark word or even phrase boundaries, and so these must be inferred by the reader. There is no 'natural' correspondence between listening and reading in representational units at the sensory level. Beyond this, however, all forms of writing do contain words, and word recognition has formed the basis of investigation of both reading and listening to language (*Fig. 1*). Words are taken as the

HERES TO PANDS PEN DASOCI

ALHOU RINHAR M LESSMIRT

HAND FUNLET FRIENDS

HIPRE IGN BEJ USTAN DKI

NDAN DEVILS PEAKO FNO NE

Fig. 1.   A sign seen in a Devon pub illustrating the importance of word boundaries in reading written English.

basic unit of processing because of the **word superiority** effect, first noted by Cattell over a century ago, which shows that letters can be identified faster when they are part of words (and even readable non-words) than when they are presented alone.

Techniques

Early research in reading used a technique in which words were presented one at a time, paced by the reader pressing a button. The duration between button presses was taken as a measure of the time taken to read each word, with difficulties in processing leading to longer durations. Unfortunately people tend to automate the button processing to produce a steady rate of presentation. A variant called **rapid serial visual presentation** (RSVP) takes the control away from the reader and the words are flashed up at a fast but steady rate; studies have found that short sections of text can be understood at presentation rates far faster than normal reading speeds (up to 1200 words a minute, compared with around 300 a minute for normal reading speeds).

A series of studies using a **moving window**, in which text either side of the word being read is made visible, showed that the surrounding words were important in reading: words were not recognized in isolation during normal reading. The words that had just been read influenced reading speed, if they could still be seen; so did words that had not yet been read, presumably because their shape and size gave some information that helped with syntactic parsing (e.g. if it was a short function word, or started with a capital letter). Improvements in technology eventually made it possible to measure eye movements and ascertain gaze direction with sufficient temporal and spatial accuracy to tell exactly where on a page a person looked and for how long.

Eye movements

Our eyes do not move smoothly, but in a series of small jumps called **saccades**. In reading, each saccade lasts about 25 msec, and moves the point of foveal fixation (see Topic B1) seven to nine letters forward at once. During a fixation, we can take in and process visual information from roughly four letters to the left of the fixation point and fourteen letters to the right. The fixations can last from 20 to 500 msec, depending on what is being read, but are on average around 225 msec. A skilled reader will make about 250 fixations a minute, reading around 300 words. Some of the saccades are regressions, going backwards, and longer and rarer words produce longer fixations than shorter and more frequent words. Not all words are fixated: readers tend to fixate on 80% of content words, and to skip over many of the short function words. Speed readers learn to minimize regressions, to decrease the number of long fixations, and to increase the size of their saccades, but still process all of the words. While their reading speed improves, comprehension of details and nuances in the text declines (Just *et al.*, 1982).

A key feature of fixations is that they do not land randomly within the text, but tend to be located between the beginning and middle of words (Rayner, 1979). This implies that there must be some cognitive control over their direction, because for them to land so systematically within words indicates that some information about the words has already been extracted during the previous fixation. Rayner and Pollatsek (1981) used a moving window experiment to randomly vary the amount of text that could be seen to the right of each fixation, and found that saccades were longer when more text could be seen. McConkie's (1979) model of eye movements during reading used the spotlight analogy of attention (see Topic C2). The attentional spotlight scans forward within the text until a difficult or degraded piece of text is encountered, and the eyes then move to fixate there to provide a richer foveal input. Although this accounts for the pattern of data, it does not match the timing of fixations: if the eye-movement system had to wait until the cognitive operations involved in reading had been unable to process the input, fixations would be far longer than they are.

Morrison (1984) modified this model (*Fig. 2*) by proposing that the attentional spotlight allows each word to be read in turn, and that the successful process-

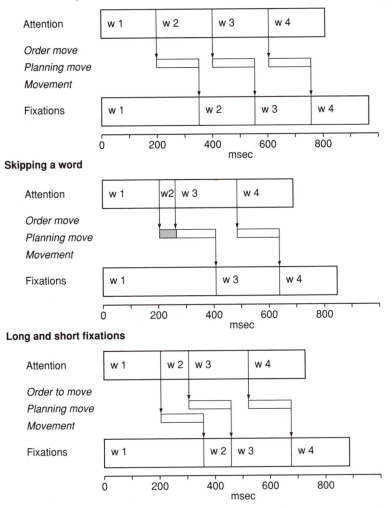

Fig. 2. *In Morrison's model of eye-movements during reading, an order to move the eye on to the next fixation point takes around 150 msec to carry out. During this time, attention moves on to the next word to start processing it (top). If it is recognized quickly, another eye movement may be ordered before the first one has occurred. If this happens within 50–75 msec, the first movement may be cancelled, and the word skipped (middle). If it happens later than this, the first movement is carried out, but the second one happens soon after, resulting in a pattern of long-short fixations (bottom).*

ing of each word results in a request to move the eyes to fixate the next word (whose position can be determined because it is partially within the current fixation). Because it takes around 150 msec to prepare and execute the eye movement, the next word might already have been processed by the attentional spotlight, and another movement request issued. Morrison's model allows this second request to cancel the first movement, provided that it occurs within 50 to 75 msec of the first request, and start the preparation of the second movement. There will thus be a moderately long fixation when more than one word is being read easily on a fixation. If the second request comes more than 75 msec after the first one, though, it is too late to cancel the first eye-movement, but the second eye-movement will occur very rapidly afterwards, within around 100 msec. There will be a pattern of long and very short fixations. Morrison's model also explains why the words that are not fixated in text tend to be the short, frequent and predictable words, because these are the ones that are very rapidly processed.

Word recognition Although there are many models of word recognition, all share some common concepts. The link between the perceptual representation of words from either heard or written language and their meaning is mediated by the **mental lexicon**. This can be thought of as a set of pointers, with a pointer from each word-form to a concept in semantic memory. Once word units have been identified in the sensory data, items in the mental lexicon that are close perceptual matches and which make contextual sense are activated, a process known as **lexical access**. In some models, only one item in the lexicon can be activated, but in others, many candidates may be activated, and further processing is required to choose between them. When there is only one active item, **word recognition** has been achieved. If a non-word has been presented it has to be rejected and not recognized as a word.

A number of well-established findings form the basis for all models of word recognition. As well as the word-superiority effect, commonly used words are recognized faster than rarer words (the **word frequency effect**), and the closer that a non-word is to a real word, the harder it is to reject in a lexical decision experiment (where you just have to decide if a string of letters is a word or not). Letter strings that do not follow the **orthographic rules** (i.e. which use letter combinations that do not occur in the written form of the language, such as **ritjq** in English) can be rejected faster than real words can be accepted, so fast that the mental lexicon is probably not involved, but only perceptual features are used. Context is also important, because words in context (e.g. 'nurse – doctor') are recognized faster than words out of context (e.g. 'nurse – lawyer'; Meyer and Schvaneveldt, 1971; see Topic D3), and incongruous contexts actually inhibit word recognition (Antos, 1979). The effects of orthography and context cannot be accounted for by entirely bottom-up models such as Selfridge's 'Pandemonium' model (see Topic B5).

Direct access models Morton's (1969) logogen model of lexical access proposes that there is a set of cognitive units (the **logogens**), with one unit corresponding to each linguistic unit (words and morphemes). These list the perceptual features that their 'word' contains, such as 'starts with a vertical line', and each matching feature in the perceptual input (e.g. L, I, P, D, etc.) increments the activity of the relevant logogens. Each logogen has a threshold of activity, and when it is reached, it fires to indicate that its word has been potentially recognized. The more frequent a word (i.e. the more often its logogen has fired) the lower the threshold becomes, so the faster it can be recognized (and also misidentified). Logogens can be **semantically primed** by contextual information, so that even before any features have been recognized, logogens with semantic associations to material that has recently been processed will have been slightly activated (see Topic D3).

This model is a **direct access** model because it predicts that cognitive units corresponding to words are directly activated by perceptual information. Although the model was developed to account for the perception of briefly presented written words, it can conceptually be modified to model spoken language perception by including auditory features in the logogens (e.g. the 'labial plosive' component of /b/ and /p/, or the short voice-onset time of /b/ and /d/). Another direct access model, specifically developed for comprehending spoken language, is the **cohort model** proposed by Marslen-Wilson (1987). The first two phonemes that are perceived activate a cohort of all matching word units. As further phonemes are perceived, items in the cohort that do not match the input are eliminated, until only one unit remains. This allows distinctive words to be recognized before they have been completely perceived (or even spoken). Whereas the units in the logogen model compete to be recognized, the units in the cohort model compete to be eliminated.

Serial access models

Instead of directly activating word recognition units, perceptual information could be gathered up and then compared against templates for candidate words (see Topic B5). Forster's (1979) search model allows the features of the word to be compared against the stored features of what a word looks like or sounds like, in two stages. The words that are candidates for matching the input are selected on the basis of initial features, as in the cohort model, and pointers to them are placed in an **access file**. The file is ordered according to word frequency, with common words appearing early in the file, and rare words later. The entries in the access file are then compared against the full perceptual input one at a time, in serial order, and when a match is found, the pointer to the **master file** for that word is followed to activate its meaning. Because the search is serial, recognition times for common words that occur early in the access file will be faster than for the later rare words, and non-words that are similar to real words and which have a larger access file to work through will take longer to reject than less regular non-words, which will generate smaller access files. Context effects are accounted for by a simultaneous serial search through a second access file that has been generated by semantic associations with recently recognized material.

Dual access models

Although the models described above could in principle be used to model either written or spoken language by varying the feature set used as input, they are in spirit unimodal, modeling one or other form of language. It is plausible that the visual features of words are not the only factor involved, even in reading. The **phonological recoding** hypothesis proposes that letter patterns are converted to sound patterns, by a set of **grapheme–phoneme correspondence** (GPC) rules, and that reading then proceeds as if the input had been heard. Evidence in favor of this came from lexical decision studies which showed that non-words that sounded like real words (**pseudohomophones**, such as *trane*) were harder to reject than other orthographically regular non-words (e.g. *sharb*), but it was later found that the pseudohomophones also looked more like real words, and so the feature detection models could account for the effects. Other data (Meyer *et al.*, 1974) suggested that sound was important, though: lexical decision times for pairs of visually similar rhyming words (e.g. *net bet)* were faster than for non-rhymes (e.g. *cow mow*).

Kleiman (1975) asked volunteers to do reading tasks while having to speak at the same time (i.e. by repeating digit names that were read out to them), and found that although rhyming judgments were affected, visual similarity and semantic relatedness judgments were not affected. He argued that although phonological recoding was especially important in learning to read, for skilled readers the primary route was through visual feature analysis. Skilled readers only relied upon phonemic recoding when they encountered new words, but low ability readers who had not developed visual feature analysis skills continued to rely upon phonemic recoding. His model therefore contains two routes (phonological recoding and visual feature analysis) and is a **dual access** model. It has been argued that difficulties with the formation of GPC rules might be one cause of reading disorders such as dyslexia (see Topic I4).

An influential theory of speech perception that is related to dual access models of reading is Liberman's **motor theory** (e.g. Liberman and Mattingley, 1985), which argues that we understand speech by reproducing in our minds the articulatory actions that we would have to make to produce the sounds that we are hearing (but only mentally; we do not actually need to make the motor movements). An advantage of this theory is that it allows slurred or slightly mispronounced speech to be comprehended because we can 'feel' what the speaker intended to say. It also explains the similar categorical perception of sounds such as the phonemes /di/ and /du/, which are produced by very similar mouth movements yet which have very different auditory features. However, Kleiman's data make a strong form of the theory implausible because the processes and also the muscles involved in speaking were being used while his volunteers were reading.

12 UNDERSTANDING SENTENCES

Keynotes

Garden path effects
Some sentences start off with word sequences that could belong to different phrase structures. When we choose the wrong interpretation, we have been 'led down the garden path' by two heuristics: minimal attachment leads us to choose the simpler of two alternative phrase structures; late closure leads us to add words to the current structure rather than closing it to start a new phrase. Garden path effects do depend upon the semantic plausibility of the alternative phrases as well as syntax, so there must be some comprehension of meaning during parsing.

Ambiguous sentences
Text is normally read in the context of previous phrases and in a meaningful situation, both of which can provide pragmatic constraints on the interpretation of ambiguous sentences such as 'put the block in the box on the table'. The fact that the syntactically most simple interpretation is not always the one that seems most obvious also shows that the semantic level must influence parsing.

Inference
To resolve ambiguities and references such as 'it', we need to construct inferences about meaning. Johnson-Laird's mental models approach argues that a reader constructs a representation of the scene being described as they read or hear a sentence, and that as each part of a phrase is encountered the new information is added to the existing model in the simplest way. Kintsch's construction–integration theory predicts that propositions that are central to the narrative will be retained in working memory for longer and hence be recalled better than non-central propositions. Situation models argue that spatial details are crucial in comprehension, and that these determine the model that is built.

Related topics   Visual form perception (B4)   Dyslexia (I4)
                        Understanding words (I1)

Garden path effects
The data from eye-movement studies has been used to identify the parts of sentences that people need to spend most time processing, and so spend longer looking at. One finding is that we try to anticipate the syntactic structure of the phrases that we are reading (see Topic H1) and tend to commit ourselves to simple structures too early, before more complex alternatives can be ruled out. This leads to the phenomenon of **garden-path sentences** in which our parsing leads us down the path of inferring one meaning, only for subsequent input to be incongruent and require us to re-parse the sentence to discover the more complex alternative.

Two heuristics influence the structures we choose: minimal attachment and late closure. **Minimal attachment** requires new words to be added onto the phrase structure by positing the smallest number of additional syntactic units. For example, consider the phrase *the boy fed the rabbit*. After hearing the partial sentence *the boy fed*, which has a noun phrase (NP) followed by a verb (V), readers will assume that it is following the simplest sentence structure of noun phrase plus verb phrase (NP+VP), and that the verb will be followed by another noun phrase, saying who the boy has given some food to (because VP=V+NP). The subsequent words *the rabbit* appear to confirm this, so the reader will then be surprised to encounter the subsequent words *burped*, which cannot be added to the assumed NP+VP structure. They will have to reassess the whole sentence to discover that *fed the rabbit* is a relative clause describing *the boy*, and so is part of the first NP, and that *burped* is actually the VP. **Late closure** requires new words to be added to the phrase currently being constructed if possible, rather than finishing with the current phrase and starting to build another, so in the current example, *burped* does prompt a re-parsing rather than starting a new sentence.

Both principles can be seen as cognitively economic, because they attempt to limit the complexity of what is being processed, and to work with existing representations rather than shifting to new ones. For most sentences, they do actually work, and reading can proceed quickly because no time need be spent building new phrase structures. Intentionally constructed garden path sentences used in eye-movement studies confirm that there are longer fixations on words at points in which these two principles break down. However, there is evidence that the meaning of the words is also crucial. Garden path effects only occur strongly where the simple syntactic structure being generated is semantically plausible, and more complex structures are readily built if they are more plausible. If the example above had been *the dog fed the rabbit...* then it is not plausible that a dog would be feeding a rabbit, but it is plausible that a dog would be given rabbit to eat, and so no garden path effect would occur. Milne (1982) showed that only the second of the following possible garden path sentences actually produced any effects:

> *The table rocks during an earthquake*
> *The granite rocks during an earthquake*
> *The granite rocks were by the seashore*

This is paradoxical, because it might be thought that the meaning of a phrase cannot be found until the syntax of the phrase has been constructed. Milne's data shows that in fact the meaning is influencing the syntactic construction, so it must be at least partially available as the phrases are constructed.

Ambiguous sentences

In normal comprehension, phrases and sentences are seldom encountered in isolation. The surrounding text and also the situation within which they are read usually provide extra information that can help to distinguish between different interpretations, or can make complicated phrase structures more pragmatically plausible. Winograd (1972) pointed out that a command like *Put the block in the box on the table* is ambiguous when encountered out of context, but unambiguous when encountered in the presence of a box and a block that have some existing spatial configuration. Partial sentences such as:

> *People living near airports know that landing planes...*
> *People trained as pilots know that landing planes...*
> *People trained as pilots know that approaching storms...*

can all potentially be continued with either *is* or *are*, but the most likely pragmatic continuations of the first two examples are *are* and *is*, respectively. The
third example is truly ambiguous, for *approaching* could be the activity that
pilots know about or the behavior of the storms (Garnham, 1985). There do
seem to be at least three levels of processing in comprehension: the word level,
the phrase level (syntactic level) and the message level (semantic level). The
way that these interact is not yet understood, but the fact that the syntactically
simplest syntactic structure is not always chosen, shows that the semantic level
must influence parsing in some way.

Inference

The way that semantic meaning influences comprehension of ambiguous
sentences suggests that the propositional information provided by previous
sentences is active within memory and influences subsequent processing. Both
written and spoken speech often make use of **anaphora**, in which a previously
referred to thing is not repeated, but replaced by a pronoun such as *he, her, it,
this* or *that*:

> *The author thought of an old friend. He was tall and handsome...*
> *The author thought of an old friend. He was writing a novel...*

In these sentences, there is no difficulty in working out that *He* refers to the
friend in the first case and the author in the second case, although the opposite
assignments could also be true. In the **distance effect** (Clark and Sengul, 1979),
the time taken to read a sentence containing an anaphoric reference increases
the further the reference is from the thing it refers to (the referent), so the
second example above should take longer to read. Simple textual distance is not
the whole explanation, though, for Clifton and Ferreira (1987) showed that it
could be abolished if the referent was still the conceptual topic of the sentence
at the point that the reference was encountered. This suggests that the active
representation in working memory of the situation being described is influencing the resolution of anaphora.

Johnson-Laird's mental models account of reasoning (see Topic G2) has been
applied to the comprehension of text to account for anaphora and the influence
of partially parsed phrases in garden path effects. The idea is that as parts of a
phrase are encountered, the reader builds a mental representation of the situation being described. Additional propositions in the text are then added to this
model wherever possible, in the way that requires the least restructuring of the
model. This produces the same results as the minimal attachment and late
closure heuristics, and explains why the semantic meaning of sentences seems
to be available before they have been completely parsed. Ambiguous sentences
which lead to two interpretations require the reader to construct two models at
the same time.

A similar idea lies behind Kintsch's (1988) **construction integration theory**,
which says that as phrases are parsed, their meaning is held in memory as a
series of logical structures called **propositions**, rather than the text being
retained. This is why the gist of text is recalled rather than the actual surface
structure of the text (Gernsbacher, 1985). Propositions (see Topic E1) consist of a
subject, which the proposition is about (also called an argument or a topic) and
a **predicate**, which describes the subject. In the sentence *the boy fed the rabbit
burped*, there are two propositions which overlap: (i) *the boy was fed with the
rabbit* and (ii) *the boy burped*. When correctly parsed, these two propositions are
easy to combine because they share a common subject. In Kintsch's theory, the

reader constructs propositional representations of the text to comprehend each sentence and then integrates these with their understanding of the text that has gone before (i.e. their ongoing interpretation of a narrative).

The propositions that are currently being read, together with the current model of the narrative, are held in working memory (see Topic D2), which is where the integration occurs. Because working memory is limited, propositions that are not central to the thematic structure of the narrative will be forgotten to make room for more meaningful material. The thematically important propositions will remain in working memory for longer, leading to better recall.

The representational nature of mental models comes to the fore in **situation models** of narrative comprehension. These argue that spatial relationships between objects are key attributes in determining what interactions can take place between them, and so the comprehension and recall of spatial information is a key component of narrative comprehension. When understanding a narrative, we build up a spatially based model of the situation that is being described (similar to Johnson-Laird's idea of mental models, see Topic G2), and our comprehension of the narrative is influenced by the spatial proximity of items in the model. Glenberg *et al.* (1987) showed that people read 'John *put on* his sweatshirt and went jogging' faster than 'John *took off* his sweatshirt and went jogging', and also were more able to later say what John had been wearing. Curiel and Radvansky (2002) constructed a story in which objects were located in different rooms of a house, and found that reading speed was affected by the notional distance between objects (similarly to Kosslyn's findings on the use of mental images, see Topic E2). Understanding anaphora was fastest when the sentence referred to an object that was in the same room as the protagonist of the story, and reading times increased when a change in spatial location occurred, suggesting that people were updating their situation models.

13 LANGUAGE AND THOUGHT

Keynotes

Relationship between language and thought	Piaget saw mental development as the result of a progressive set of developmental steps, with the initial development of thought logically preceding language development. Whorf argued the opposite, that language provided the categorical boundaries needed to classify and interpret the world. Vygotsky reasoned that infants without language can clearly do some thinking, but that as language develops it has more and more influence over a child's environment and hence their experience of the world.
Linguistic universals	Chomsky proposed that the syntactic structures shared by different languages showed the existence of an innate ability to learn language and a 'universal grammar'. In most languages subjects precede objects, for example. Good words can be made bad by adding a negative morpheme, but rarely vice versa. The normality hypothesis explains this as being due to the fact that the common state of things is expressed by the positive, good word, and the rarer state by the negative, bad word.
Linguistic relativity	The Sapir–Whorf relativity hypothesis argues that language constrains thought: you can only think of something that your language has words for. There is some evidence for a weaker form of relativity, that language influences thought. Because language is used to express our thoughts internally, we may be able to recall the label that we gave an event rather than the actual event.
Theory of mind	Children's difficulties in tasks that are taken to show limitations in their cognitive abilities, may actually be due to them being misled by the strange testing situations and the violation of conversational principles. Altering the phrasing of the question can often improve performance. Evidence from stroke patients shows that once acquired, language and theory of mind are neurologically separate, but limitations in either when young might still influence the other one's acquisition.
Related topics	Communication (H1) Cognitive development (J1) Language development (H3)

| **Relationship between language and thought** | Although different languages have different words and different grammars, they all describe humans' interactions with each other and their surroundings. This has led to two contrasting questions: are there structural **linguistic universals** shared by all languages that can give us insight into cross-cultural similarities and the structure of the human mind; or does **linguistic relativity** mean that the way that a language allows its speakers to describe the world forces them to think of it in different ways? The views of theorists on the relationship between language and thought differs widely. For Piaget, mental development was the progressive enrichment of the ability to think, and this supported language development. A |

child had to develop the mental concept of something before they could learn the word or the grammatical construction to express it. For Whorf, all thought was expressed in language and so the language one spoke limited or shaped the thoughts one could have. Language defined the categorical distinctions between things, and so guided the conceptual distinctions one acquired. Vygotsky took an intermediate position, that they are not intrinsically linked but, being products of the same mind, will obviously influence each other. Infants who cannot yet understand language can clearly do some thinking, and need to be able to think about their environment in order to understand what their parents' speech means, so before the age of about two years thought precedes language; yet as children acquire language they use it to represent their own thoughts, and it becomes an increasingly important part of their environment.

Linguistic universals

Although languages do indeed differ, they all have things in common, as Chomsky and Hockett recognized (see Topic H1). They all contain vocally expressed words that are nouns, words that are verbs, and words that describe the nouns and verbs, as well as words for spatial dimensions, colors, shapes and kinship relations. These linguistic universals may arise because the world is pragmatically divisible into objects and actions, and human sensory physiology provides the percepts of color, shape and space that then need to be described. For Chomsky, the shared syntactical aspects of language, or the **universal grammar**, indicated that those aspects of syntax were also in some way innate, or that the processes for learning them were innate. An example of a syntactic universal is the placement of the subject of a sentence before the object in active sentences (although the location of the verb can vary). A very small number of languages use the order verb–object–subject, but no languages place the object first in the sentence. Other universals are less obviously environmental or innate, but seem more cognitive. These include number, causation, negation and time.

An example of a linguistic universal comes from color words. Cross-cultural studies have found that languages contain a variety of different color names. A small number of languages only have two color words, corresponding to dark and bright (i.e. black and white), while others have three to ten basic color words. Berlin and Kay (1969) found that when a language has three color names, they are always black, white and red; languages with four or more words also include one or more of yellow, green and blue; with seven or more they add brown; and with eight or more they include one or more from violet, pink, orange and gray. The order of increasing complexity of color naming is relatively consistent, suggesting that however a language divides up color, the way that we all perceive color is constant across cultures (*Fig. 1*).

Negation is a good example of a cognitive universal. Describing the attributes that a thing has is easier than describing all of the attributes that it does not have, and so languages tend to have a simple way of saying positive facts about something, and a more complex way of saying negative facts about something. In English, for example, we can say *the dog is black*, but to express the negative

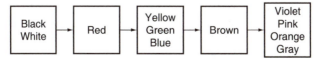

Fig. 1. The order of inclusion of color names in languages. Words for all colors within a box are included in a language before any from the next box are added.

have to add the morpheme NOT, as in *the dog is **not** black*. The relationship between goodness and positive expressions is also universal: in many languages *bad* is expressed as *ungood* or *not-good*. It is common to form a 'bad' word from a 'good' word by negating the good word (e.g. the opposite of *happy* is *unhappy*, and that of *safe* is *unsafe*) but very rare to be able to form a good word from a bad word by negation (e.g. the opposite of *sad* is not *unsad*, nor is *undangerous* the opposite of *dangerous*). One explanation, known as the **Pollyanna hypothesis**, is that people are disposed to perceive the world in a happy, rose-tinted light and so have words primarily for the good things. Bierwisch (1967) pointed out that the good terms often describe the usual state of affairs, and the bad terms unusual states. This **normality hypothesis** also accounts for pairs of adjectives that are marked in the same way but which are not necessarily good/bad, such as *usual* and *unusual* themselves, and also for unnegated dimensional terms like *tall–short* where things are said to be some amount tall, rather than some amount short.

Linguistic relativity

The linguistic relativity hypothesis is often associated with Sapir and Whorf, who argued it most strongly. The Sapir–Whorf position is that the structure of language determines the structure of thought; that thought is essentially constrained by language, and you cannot think easily of something that you do not have words for. There is no doubt that some languages have many different ways of describing something that is common and important in their culture, while another language from a culture where that thing is rare or unimportant will usually only have one or two words for it. Most people have heard the assertion that the Inuit have a large number of different words for snow, and can therefore perceive snow in a different, and more detailed way than can an English speaker. In fact, this is not true, because there are only four basic words for snow in Inuit. The Inuit languages chunk nouns and adjectives together so that a single word might mean 'wet snow', another word might mean 'slushy snow', and another 'driving snow'. As you can see, English speakers can easily conceive of many different sorts of snow, despite having just one basic word for it. Inuit people might be able to discriminate many additional forms of snow, but then they will have encountered snow more often. In this case at least, the environment is shaping both language and thought.

The strong version of the Sapir–Whorf hypothesis is not sustainable, but there is evidence that a weaker version might be: rather than limiting or constraining thought, language influences thought (Hunt and Agnoli, 1991). This follows from the evidence that we do not necessarily think in well-formed, grammatical **internal speech** but in a propositional form that has been called **mentalese** (see Topic E1). A range of views are possible on the similarity between one's mentalese and language: Fodor (1976) argued that they would be very similar; Schank (1972) that they were unrelated, and that speakers of different languages shared the same mentalese. Although thought need not be linguistic in form, expressing a thought verbally does provide another way of holding it in mind and rehearsing it (**phonological recoding**). Later on, we may be able to remember the verbal label rather than the original thought. This would provide a way that linguistic constraints could differentially favor some thoughts or discriminations over others, or facilitate some mental operations required to comprehend a situation.

Evidence that language does affect behavior comes from a variety of sources. Although color naming schemes have been used as evidence of linguistic universals, and people from different cultures show no difference in color discrimination tasks, speakers of color-rich languages do perform better on color memory tasks than speakers of color-poor languages (Stefflre *et al.*, 1966). The Navajo language emphasizes the shape of objects, and children brought up to speak Navajo sort objects according to shape, whereas English-speaking Navajo children tend to sort them according to color (Carroll and Casagrande, 1958). In a social cognition experiment, bilingual Chinese and English speakers were given a description of a person that matched a stereotype that could be expressed by a single expression in Chinese, but not in English (a respectful, wise, polite, family-oriented person, which has a simple name in Chinese) or vice versa (e.g. *bohemian*, which does not have a corresponding label in Chinese), and then asked to recall the person in their own words. If they had read the description in the language that contained the easily labeled stereotype term, their recollections conformed to the stereotype, but if they had read it in the other language, they were less influenced by the stereotype (Hoffman *et al.*, 1986).

Theory of mind   In developmental psychology, Piaget's stage theory of development (see Topic J1) is based upon the systematic types of error that children make when faced with certain tasks. It has been criticized because the situations that the child is placed in are often unusual, and the questions are phrased in complicated ways. **Conservation tasks**, for example, involve a child correctly identifying which of two identical containers holds the most liquid (for example), but then incorrectly switching their choice when the liquid in that container is poured into a differently shaped container (*Fig. 2*). In this case, it may be the repeated questioning that is influencing the child, because normally when an adult asks a child something twice it is because they got the answer wrong the first time, not because they got it right.

Similarly, egocentrism and **false belief tasks** suggest that a child is apparently unable to see things from another person's (or toy's) point of view. These have been used to argue that young children do not have a fully formed theory of mind, but believe that anything they know is also known by everyone else. In the Naughty Teddy task, for example, an adult shows the child two teddies putting a ball in one container, and one then leaves. The remaining 'naughty' teddy moves the ball to another container. When asked 'where will the other teddy look for the ball?', most 3-year-old children fail to point to the first container, but point to the second container which does actually contain it. Modifying the question slightly, to ask 'where will the other teddy look **first** for the ball?', allows 3-year-olds to perform mostly correctly. This suggests that it is the phrasing of the question that is misleading them, because they assume that it means where the teddy will **have** to look to find the ball. The questions in these tasks are confusing the children because they are not conversationally explicit: they break Grice's maxims of co-operation (see Topic H2) by not alerting the children to the fact that this is an unusual sort of thing to be asking, and that they should not make inferences based on their normal experiences of being asked questions.

The direction of causation in theory of mind problems and language disorders is complex: an underlying inability to develop language (or the lack of linguistic input caused by deafness) might prevent a child from interacting socially, and hence prevent them from understanding other minds; or an underlying inability to interact socially might be the root cause of autism, and by

Fig. 2.   An example of a conservation task. Children shown the top row of three beakers will readily say that the middle beaker contains the most drink. After seeing the drink poured into the left hand beaker, though, they will say that the right hand beaker contains the most drink, because the level is higher.

limiting interaction prevent the child experiencing language richly, so preventing them from performing well on complex, counterfactual questions. Adults who had a normal childhood and developed full language abilities, but who have suffered strokes later in life, provide evidence that, once acquired, language ability and the ability to infer other people's knowledge are neurologically separate. Patients with right hemisphere damage, whose language abilities are unimpaired, perform poorly on false belief tasks, while those with left hemisphere damage causing aphasia perform well (Siegal *et al.*, 2001).

14 DYSLEXIA

Keynotes

Reading difficulties	People who have learnt to read normally sometimes report difficulties in reading following brain damage (acquired dyslexia); other people have a difficulty in learning to read in the first place (developmental dyslexia), although dyslexia is not the same as poor reading ability. The specific patterns of disability in acquired dyslexia help to identify the component processes involved in normal skilled reading. Developmental dyslexics often develop strategies to overcome their problems and do not realize that they are dyslexic until adulthood, when it may be too late to help them.
Acquired dyslexia	Three main patterns of dyslexia following brain damage are deep dyslexia (inability to read words letter-by-letter, but access to word meanings via word shape); surface dyslexia (inability to recognize whole words, but must spell a word out letter-by-letter to 'hear' it); visual dyslexia (confusion between words containing similar letter features).
Developmental dyslexia	Reading requires a large number of cognitive skills to be learnt, so difficulties in any one of them can impair the whole process. Bright children can often compensate by developing strategies to overcome a single deficit, and so it can be difficult to identify their problems. Although it was once thought that poor eye movements might be the primary cause, it now seems that reading difficulties could be causing them instead.
Processing deficit	Some theorists argue that a deficit in visual processing makes it difficult for letters to be recognized, while others argue that all rapid processing of stimuli is a problem. The most prominent hypothesis is that dyslexics have some phonological deficit that prevents them learning grapheme–phoneme correspondence rules. This slows down their reading and prevents them acquiring the whole-word recognition skills of skilled readers.
Cerebellar deficit	The cerebellum is a region of the brain that learns to carry out motor actions automatically. The cerebellar deficit hypothesis argues that a general deficit in automating the component tasks of reading would lead to all of the specific processing deficits that have been proposed as causes of dyslexia. It has been shown that dyslexic children have difficulty performing cerebellar tasks, such as balancing on an unstable platform, if they are given an attentionally demanding task to do at the same time.
Related topics	Sub-disciplines of cognitive psychology (A4) Understanding sentences (I2)

Reading difficulties   Dyslexia is defined as a difficulty in reading despite an otherwise normal range of cognitive skills. It describes the observable outcome of a wide range of underlying problems, in which there is a disturbance in one of the component processes involved in reading, either caused by brain injury after they had learnt to read normally (**acquired dyslexia**) or by a problem in learning to read in the first place (**developmental dyslexia**). People who are truly dyslexic are not simply poor readers, nor are poor readers necessarily dyslexic, because there will always be a range of achievement levels on a skilled, multifactorial behavior like reading. Their problems can usually be attributed to one of several patterns of specific reading difficulty, and this evidence has helped in the development of a general model of skilled reading. Acquired dyslexia provides the clearest evidence about the component processes of reading, because we can be sure that the individuals who have lost the ability to read did have all of the necessary abilities before their injury or stroke. The specific difficulties that they have following their injury can be presumed to reflect damage to one or more of the component processes. Developmental dyslexia is more problematic, because people develop different strategies to cope with their problems, and sometimes reading difficulties are revealed under stressful circumstances, when they are unable to concentrate fully or need to read very quickly. Although developmental dyslexias can be moderated by training if they are identified early enough, the chicken and egg problem of needing to identify a difficulty in reading before reading has been learnt means that the problems of many bright dyslexics are not identified until adulthood; indeed it is most likely that bright children will find ways of coping with their dyslexia until it is too late to help them.

Acquired dyslexia

Three common patterns of acquired dyslexia were identified by Marshall and Newcombe (1973), and a number of rarer forms have been described since. The first two forms are named by the abilities that are left intact. People with **deep dyslexia** find it hard to recognize words from their letters, but can recognize whole word forms, and can often retrieve appropriate semantic information even when they cannot recognize the word. This means that when shown the word *bungalow* they might reply *maisonette*, a very dissimilar word in its surface structure, but in the same semantic category of small dwellings. Because they apparently have a difficulty in using the surface structure of words, they can make confusions between different words that look alike, such as *violin* and *violet*, and can misinterpret the grammatical endings of words (inflections) as in *jumper*, *jumping* and *jumped*. Deep dyslexics can read text containing common words reasonably quickly, but still have difficulties with grammatical function words which have little semantic content.

People with **surface dyslexia** have the opposite pattern to deep dyslexics, and cannot access the semantic meanings of words from their written form, but have to spell them out to themselves, letter by letter, as if this allows them to hear the words. This leads them to make errors on words which do not have a straightforward pronunciation (a problem which is especially noticeable in English), such as *pigeon*, and to confuse words whose spelling leads to similar pronunciations, such as *haven*, *heaven*, *heaving* and *having*. Taken together, these two patterns of dyslexia suggest that in normal skilled reading, people possess two routes for recognizing words, one based on the direct activation of semantic information from the overall shape of the word, and one based on an internal, auditory representation derived from the graphemic information (the shapes of letters and letter clusters). It is noticeable that in both sets of dyslexics, semantic context still plays a large role, with words in text being identified more accurately than words in isolation.

The third form of dyslexia, **visual dyslexia**, reflects a problem in recognizing the visual form of words. While these dyslexics can name all of the individual letters in a word correctly, they make errors in identifying the word which suggest that one or more of the letters have been misread (e.g. *C, A, B = cap*). They are not, as some have thought, having difficulty in identifying the orientation of letters, because they can name them all correctly. They may be having difficulties in **visual segmentation**, in that the features of one letter or word can overlap with another, or similar features can be confused with each other. This could come about through a lowered threshold for word recognition, with common words being incorrectly reported on the basis of insufficient visual information.

Developmental dyslexia

From the account of reading presented throughout this section, it is clear that a large number of basic perceptual and cognitive skills are involved. If a child has a problem in developing any one of these skills, then it may also prevent the development of subsequent cognitive skills that rely upon its correct functioning to provide them with information. In consequence, a number of hypotheses have been advanced to account for the developmental dyslexia, and all may have some validity, but none are likely to be the sole explanation.

An early suggestion was that the control of **eye movements** could be at fault. While it is true that the pattern of saccades across text in dyslexics is different to that in normal readers, this could be caused by their reading problems, rather than causing them. If they are having difficulties understanding the text, they will need to make more regressions, and if they cannot read whole word shapes, they will need to look in more detail at individual letters. Pavlidis (1985) presented data that showed some dyslexics also had poor eye movement control on non-reading visual tasks, especially a difficulty in maintaining fixation and a tendency to show more right-to-left saccades than left-to-right saccades. Others have failed to replicate this, and have also found similar problems in children without reading difficulties, so while it may be a cause of some dyslexics' problems, it can be neither a sufficient nor necessary cause. Stein and Fowler (1993) found that some dyslexics had difficulties in vergence control, keeping both eyes pointing inward and focused upon the object being attended to, and that six months spent wearing spectacles that improved vergence also improved reading skills.

Processing deficit

A number of hypotheses center on the idea that after visual sensory information has been received, there is some problem in the basic **perceptual processes** by which text is recognized. These include visual stress caused by flicker as the eye moves across the black and white lines of text (Wilkins, 1995), prolonged persistence of visual images causing interference across saccades (Lovegrove, 1994), and slow processing of visual input (DiLollo *et al.*, 1983). Some have argued that dyslexics have a difficulty processing any perceptual input rapidly, whether or not it is visual (a **rapid temporal processing deficit**). Tallal *et al.* (1993) reported that dyslexic children found it difficult to say which of two tones came first when they were presented rapidly in succession. All of these hypotheses are controversial, however, and there is as much negative evidence as evidence in support.

The most prominent hypothesis has been that there is a **phonological deficit** making it difficult for dyslexics to develop grapheme–phoneme correspondences, and hence also slowing down their ability to learn word shapes (e.g.

Snowling, 1987). They thus show the same problems in reading words letter-by-letter as deep dyslexics, without having had the benefit of years of normal reading skill to provide them with a direct word-level access route to semantic meaning. Fawcett and Nicolson (1995) found that three dyslexic groups aged 8, 13 and 17 years all performed less well on tests of sound categorization than both their chronological and their reading age controls. This hypothesis also accounts for the finding that dyslexics do poorly on tests of auditory short-term memory (see Topic D2). Shankweiler *et al.* (1979) found that dyslexics made fewer phonological confusion errors on short-term memory tasks, suggesting that they made less use of auditory rehearsal. Neurologically, Frith (1997) argues that an abnormality in the perisylvian region of the brain can cause a difficulty processing sound, leading to poorer grapheme–phoneme correspondence skills.

Cerebellar deficit The **cerebellar deficit** hypothesis brings together many aspects of the other hypotheses and proposes that the basic deficit in dyslexia is one of the learning of skilled behavior, namely the automatization of motor skills. In reading, many different perceptual and cognitive skills have to be performed automatically, without conscious attentional control. The transition from a controlled skill to an automatized skill is attributed to the cerebellum learning to carry out motor actions directly from sensory input, without requiring higher cognitive effort:

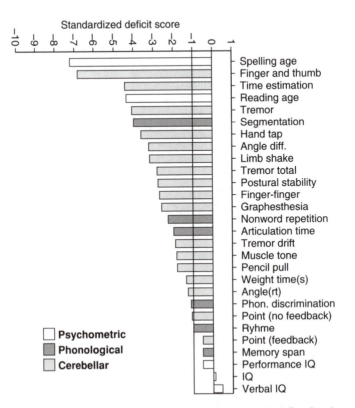

Fig. 1.   Children identified as dyslexic due to a deficit in their Reading Age also show deficits in a wide range of other tasks thought to require cerebellar activity because they involve the automatization of skills. This suggests that dyslexia may also be a problem in automatizing the skills involved in rapid reading. (data from Fawcett, Nicolson and Dean, 1996)

this leaves more capacity to be directed towards comprehension and inference. The strongest support for this hypothesis comes from dyslexic children's ability to balance while doing various mental tasks. Nicolson and Fawcett (1990) found that when not distracted by an additional task, dyslexic children could balance as well as other children, but when asked to count backwards at the same time, they tended to wobble and fall. The hypothesis is that all of the specific difficulties in eye movements, visual processing and phonological processing arise from this underlying difficulty in automatization. It also predicts that dyslexic children should show deficits in a wide range of 'primary' skills such as bead threading, choice reaction time, color naming and non-word repetition, as indeed is the case (*Fig. 1*, Fawcett, Nicolson and Dean, 1996).

J1 COGNITIVE DEVELOPMENT

Keynotes

Neural development

The brain of a human newborn baby is very immature. Although it contains almost all of the neurons that it will ever have, they have very few interconnections. Growth of these connections is very rapid, leading the brain to reach 75% of its adult weight by 30 months. After this, weaker connections die off in a process called pruning, leading to increasing specialization of brain regions. Neural function is fairly plastic, and practicing a particular cognitive skill enlarges the region of the brain that governs it. As we age, loss of neurons and connections weakens memory, concentration and motor control.

Piaget

Piaget's influential theory of cognitive development is a stage-based account, which recognizes that the infant mind is immature and needs to mature neurally to support development. In the sensorimotor stage, the newborn child develops action schemas through a process of equilibration with its environment, assimilating novel objects to existing schemas and accommodating existing schemas to new situations. In the pre-operational stage, the two-year-old child combines its schemas into rules for manipulating their environment. In the concrete operational stage, seven-year-old children can form goals and know how to reach them, but do not always generalize their schemas to new situations. This requires formal operational thought, which is reached by the age of eleven.

Vygotsky

Vygotsky argued that the child's interaction with the environment, especially their social interaction, was much more important than neural maturation. Vygostky saw development as continuous, not happening in discrete stages, with progress occurring rapidly in familiar domains and slowly in domains that children had little exposure to. The gap between a child's performance (the things that the child does) and their competence (the things they are potentially able to do) is their zone of proximal development. In this zone lie tasks that they can learn to do if challenged.

Information processing

The maturation and development of cognitive skills allows faster and more efficient strategies for information processing. This in turn allows a richer form of experience of the world, accelerating cognitive development. Children learn metacognitive skills to select ways of encoding information, of rehearsing it in short-term memory, and of monitoring their own performance, to select the best strategy for a task. Case argues that this knowledge is organized in domain-specific conceptual structures that support learning across tasks within each domain. More efficient use of the phonological loop of working memory, for example, supports language acquisition; better language abilities support efficient use of the phonological loop.

Related topics

Short term memory (D2)
Connectionism (E4)

Language and thought (I3)

Neural development

Like many newborn mammals, human infants are quite helpless at birth. Most mammals mature rapidly, learning to stand, walk and explore their environment within a few days of birth. Humans remain helpless for much longer. This is partly due to the size of our brains (and heads) compared to the size of the female pelvis: we have to be born with immature brains, or our mothers would not be able to give birth to us. Thereafter the cortex develops rapidly, and continues growing in spurts until adulthood. Although a neonatal child has almost all of the brain cells (neurons) that it will ever have, they have very few interconnections. Neural development is almost entirely the growth, and then refinement, of connections between neurons: a 1-year-old has 10 times the connections of a neonate, and they continue increasing until the child is about 2 years old. After this, the brain continues to increase in size more slowly, but the number of connections decreases. A 30-month-old child has three-quarters of the brain capacity of an adult, but twice as many neural connections. The adult's remaining connections, however, have become larger and stronger, with thicker myelin sheathing, to allow the electrical activity to travel further through the brain. The refinement is thought to reflect the removal of connections that are not used as much, allowing an increasing specialization of function of specific areas of the brain, and for this reason has also been called **pruning**. It also means that neural development is somewhat plastic: cognitive skills that are practiced a lot recruit richly connected neural areas, while brain areas supporting cognitive skills that are not practiced diminish. In violinists, for example, more of the somatosensory region of the cortex is activated by stimulation of the fingertips, and the size of this particular region is related to the age at which they started to play their instrument (Schlaug et al., 1995). Even in older brains, the functions supported by areas that have been damaged can be taken up by remaining areas, although this is more likely to happen the younger that the brain is. Incidentally, the well-known 'fact' that we only use ten per cent of our brains is completely untrue. As a normal brain ages, the loss of neurons starts to have an effect upon its owner's abilities. Small strokes and an accumulating loss of connections can weaken memory, concentration and motor control, leading to dementias.

Piaget

The development and refinement of a child's neural capacity is reflected in the progressive acquisition and development of physical and cognitive abilities. The most influential theory of the development of cognitive and intellectual skills is that proposed by Piaget. He argued that adult capacities in these areas have to be built upon earlier stages, and that like a brick wall, these have to be built, piece by piece, from the ground up. He followed in the empiricist tradition of philosophers like the seventeenth-century John Locke, who argued that the mind of a newborn was a blank slate (*tabula rasa*), to be written upon by experience with the world. The early empiricists had thought that a child's mind was essentially adult-like in its capabilities, and it was just lacking in content. Piaget rejected this learning-based account of development. He recognized that the infant brain was immature, and that it was not just content that was added as it grew, but that the infant mind was not capable of reasoning in an adult fashion. Piaget's theory was that a child learns to order and structure the world through its exploration and interactions in its environment, and this learning is instrumental in guiding neural development. Initially, a child can only think about what is immediately present and accessible to it through its senses; as it matures it acquires increasing abilities to think about what is no longer perceivable, and to reason symbolically about abstract things.

Piaget's theory is a **stage theory** of development. From birth until about two years old, the child's behavior is dominated by physical exploration and experimentation, in the development of its **sensory–motor intelligence**. They learn that sensations from the world can be attributed to the existence of objects, and discover **object permanence**, realizing that things in the world have an existence that continues even when they can no longer be seen, heard or touched. As they learn about more and more things that they can do, their **schemas** for interaction both **assimilate** new objects (i.e. they learn that something they can do can also be done to this new object) and **accommodate** to new situations (i.e. their schemas change slightly to allow something they can do to be done in a slightly different, and more general, way). This process of extending their schematic knowledge to keep pace with the challenges offered to them by the world is called **equilibration**.

By the age of two, the child begins to be able to think of objects that are not present; not just objects that have been present and ceased to be perceivable, but things that they can recall from memory. This marks the start of **representational thought**, a qualitative change in a child's intellectual abilities that leads on to the second stage, the **pre-operational period**. This stage lasts for about five years, as the child gradually learns rules (called **operations**) about the relationships between different action schemas, and about the before-and-after states that lead to and result from the use of schemas. By the age of around seven, a child will have reached the **concrete operational stage**, having learnt that to move from one particular state of the world to another, they can **operate** upon the world by applying a particular schema: they can form goals and know how to reach them. Their behavior is still limited, though, by the surface qualities of situations, because changes in certain attributes of the situation may not lead them to recognize the changes in their actions that need to be performed to reach their goal. The final **formal operational** stage is reached when the child is able to replace the concrete nature of the world by abstract or symbolic representations that can allow a single schema to be applied in many different but related situations, and even to be applied entirely mentally without reference to any real-world objects or situations.

Some theorists, known as **neo-Piagetians**, have argued that a fifth stage of development is also possible in adulthood. Following formal operational thought, adults can learn to deal with ambiguous situations, realizing that sometimes there is no one true answer to a problem or resolution to a situation, but that incompatible representations of it have to be held in mind or taken into account simultaneously. This stage is variously called **postformal** or **dialectical** thought.

Piaget's theory influenced educational policy, encouraging people to see the explorational play of infants as necessary precursors for academic education. It is no longer accepted as entirely true, because it was largely based on experiments that purported to show that children of a particular age, and hence at a particular stage of development, made characteristic errors on certain tasks that older children could perform correctly. Later researchers have found that modifying the questions asked of children, or making the situations more familiar, resulted in behavior appropriate to later stages, throwing into question the neat ordering of development across the whole range of cognitive abilities (see Topic I3). When tested in appropriate ways, that make use of their limited motor skills, very young children have been found to have perceptual abilities such as object permanence, and number discrimination, which are denied to them by

Piaget. The general empiricist approach of exploration, discovery and progressive abstraction within domains is generally accepted, but the idea that development follows discrete, qualitatively different stages is not. Instead, development is seen as a more piecemeal, and continuous, accumulation of knowledge and strategies for the use of that knowledge.

Vygotsky

Where Piaget emphasized the role of neural maturation in the progressive stages of cognitive development, Vygotsky argued that the child's interaction with the environment was much more important, especially their social interaction. Children absorb knowledge about how to behave in particular situations and what to do with objects by watching how other people behave, and listening to what they say. Through a process called **internalization** they imagine themselves doing the same thing, and then when a similar situation arises they can try out the behavior for themselves. A key concept in Vygotsky's idea is the **zone of proximal development** (ZPD). There are things that the child can do (their established level of **performance**), and things that the child is not yet ready to be able to learn (the upper limit of their **competence)**, but in between are things that the child could do, if the opportunity to internalize and then act arose. Problems within the ZPD can be solved, but only if the child is put in a situation where they can watch other people solving them and are then encouraged or allowed to try them for themselves. Successful learning raises both the level of performance and the level of competence of the child, continually moving the ZPD onwards. Vygotsky died young, in 1934, and his approach was overshadowed by that of Piaget, who lived until 1980. His ideas have been rediscovered in recent years and have become more popular as an alternative to Piagetian psychology. The ZPD leads to the idea that testing and assessment should be designed to move children's development forward, rather than just measuring how far they have got, and this has found favor among educationalists dissatisfied with an emphasis on pigeonholing children into stages.

Information processing

Where Vygotsky looked at the social context that supported development, the information processing approach to development looks at the maturation and acquisition of the cognitive skills necessary for adult thought. An adult has developed strategies for short-term and long-term memory use, has acquired a rich body of semantic and episodic memory to help interpret stimuli, and can group and divide the world to facilitate processing information. This is not only content, as the empiricists would have argued, but also affects the way that information is processed, allowing it to be represented in many different ways, or economically coded as single chunks, and for problems to be solved through analogy (see Topics F3 and D3). In a bootstrapping fashion, increasing experience of the world allows more efficient information processing, allowing a richer experience of the world. In addition to any development in the capacity of a particular process, children may also develop better ways of using whatever capacity they currently have. They can learn to monitor their performance, to choose between different strategies for doing a task, and to use feedback about their performance so that they not only improve cognitively but also improve their **metacognitive** skills (knowledge about their cognitive abilities).

Case has argued that the abilities children have on specific tasks reflect **central conceptual structures** (CCSs), which are general but domain-specific sets of knowledge. CCSs help children think about problems in each domain, providing a basis for the acquisition of insights about problems, which further develop the CCS for that domain. Learning something about one task in a domain allows children to apply their new knowledge to other tasks in the domain. Development, in Case's model, occurs within domains as CCSs grow and merge. This view provides a link to individual difference research, and Case (2001) shows that developmental tasks reveal a set of abilities similar to those from intelligence testing (see Topic J2).

An example of the interaction between a cognitive ability and performance is the role of the phonological loop (PL) of working memory (see Topic D2) in language development. The PL allows the temporary storage and mental rehearsal of sounds, especially speech. Its storage capacity can be measured by asking children to repeat non-words with an increasing number of syllables, and finding the length at which they start to make errors. In general, it has been found that PL capacity is related to vocabulary size, grammatical competence and performance, and mean length of utterance (Baddeley et al., 1998b). The argument is that storing material in working memory allows it to be added to long-term memory, and so children with larger PL capacity can develop better long-term memory; this then allows them to chunk material more efficiently (i.e. to remember a word of several syllables as one word rather than as several syllables; which is why non-words are used to measure PL capacity) and hence increases their PL capacity. The counter-argument is that cause and effect is actually the other way around, and that language development supports the development of working memory. There is no doubt that if some domain of development is restricted for some reason, then a range of cognitive abilities reliant upon experience within that domain will be affected. Whatever the pattern in this case, however, both positions agree that difficulties in information processing are related to problems in cognitive development.

J2 INTELLIGENCE TESTING

Keynotes

Intelligence quotient (IQ)	Binet developed the measurement of mental age by giving children a graduated series of tasks. In the US his tests were modified to produce the Stanford–Binet test for use with adults, and later the Wechsler Adult Intelligence Scale, and the Wechsler Intelligence Scale for Children. These introduced the Intelligence Quotient or IQ, whereby a person's performance is compared to the average score of the whole population, to produce a normally distributed variable, one whose distribution of scores follows a symmetrical 'bell curve'. Half of people will score above 100, and half below; two-thirds will score between 85 and 115; only 2% will score below 70 or above 130. This statistical measure does not mean that intellectual ability really is distributed normally in the population.
General intelligence, g	All intelligence tests are built up of a number of sub-tests measuring different aspects of intellectual ability. The results from these tests are correlated, such that someone will usually score similarly on all of them. Factor analysis assumes that a score on a sub-test reflects two components, general intelligence, or g, plus specific ability on the sub-test. Some theorists argue that g measures a brain's neural efficiency, others argue that it is an artifact of the testing situation.
Cognitive correlates	Measures of cognitive processing that are not derived from intelligence tests also correlate with g, especially those that measure some aspect of processing speed, supporting the neural efficiency theory. These include choice reaction time, lexical access speed, and inspection time measures. Working memory capacity and the ability to control attention have also been suggested as correlates of g.
Factorial models	Thurstone felt that seven specific abilities remaining once g had been extracted were important, and reflected the fact that people were better often at one thing than another. Other factorial theories have been proposed by Guilford, whose structure of intellect model included up to 150 different capabilities, and Gardner, whose idea of multiple intelligences focuses on the skills people use in their everyday lives. Carroll related intelligence testing to information processing measures, and proposed a three-level hierarchy with g at the top and 'elementary cognitive tasks' at the bottom. In the middle were two abstract factors first described by Cattell: fluid ability (performance on novel tasks) and crystallized ability (performance on practiced tasks).
Related topics	Cognitive development (J1) Nature and nurture (J3)

Intelligence quotient (IQ)	Although all theories of cognitive development (see Topic J1) agree broadly that children develop in an orderly progression, the rate of development varies from

child to child. At the end of the nineteenth century the French educationalist Binet was asked to find a way of evaluating the developmental status of children, and in particular to distinguish mentally retarded children who would not benefit from standard academic schooling, but who needed remedial education. He devised a series of tests of scholastic ability that could be performed by normal children of a particular age. Children who were more able than their age group would be able to complete the test items suitable for children older than themselves; children who were less able would fail on items that children younger than them could complete. With his colleague Simon, he had introduced the concept of **mental age** to go alongside chronological age.

In the US, this approach was adapted at Stanford University to produce the **Stanford–Binet** test that could be applied to adults, particularly to divide army recruits into those who could be officers, and those who should remain privates. Its successors, the **Wechsler Adult Intelligence Scale** (WAIS) and the **Wechsler Intelligence Scale for Children** (WISC), introduced a way of scoring ability by comparing a person's score with the average ability score of the whole population: the intelligence quotient or, as it has become popularly known, IQ. This was necessary because after childhood, the relationship between chronological and mental age that was measured by Binet's original test becomes meaningless (it implies that a 40-year-old with the abilities of an 80-year-old would be as clever as a 4-year-old with the abilities of an 8-year-old). To standardize scores on intelligence tests, it is assumed that intelligence is **normally distributed**, that is, that there are the same number of people of above average ability as there are people of below average ability, and that the further a score is from the average, the rarer people of that ability level become. The average intelligence score is set at 100 points, and the standard deviation is set at 15 points. This means that about two-thirds of the population will have intelligence scores between 75 and 115, with only 1% scoring below 60, and 1% above 130 (see *Fig. 1*). It must be remembered, though, that this scale is an entirely artificial imposition upon the real distribution of ability, which may not be normally distributed, and might not even be a single thing.

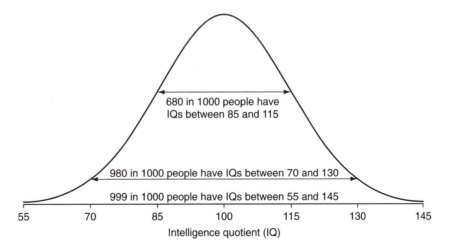

Fig. 1. Intelligence tests are designed to produce a spread of results that are 'normally distributed' with a mean of 100 and a standard deviation of 15 points. This means that 68% of people will score between 85 and 115; 98% between 70 and 130, and only 0.1% will score less than 55 or more than 145.

General intelligence, g

Since Binet's original test, measures of IQ have been based on a number of different aspects of cognitive ability. Binet included items that included testing whether a child knew how to do something, items that measured whether they knew how to adapt a known strategy to a novel situation, and items that measured the capacity for self-criticism of performance. The Stanford–Binet tests included verbal reasoning, numerical reasoning, figural reasoning and short-term memory tests. The WAIS includes a verbal scale, which contains vocabulary, analogy, arithmetic and memory tests, and a performance scale, which includes object assembly, picture completion, block matching, and digit–symbol transcription. While all of the individual components produce separate scores, and individuals might score well on one part, but below average on another, the designers of the tests felt that they were all tapping a single aspect of ability, namely intelligence. Although it is not easy to define, intelligence is thought of as a general ability to do well at all sorts of tests, and in particular, to adapt one's performance and to learn to do new things. In accordance with this, scores on the various sub-components of intelligence tests do tend to **correlate** well, that is, someone usually scores well across the whole range of tests, or averagely, or poorly (*Fig. 2*).

In the 1920s, Spearman developed a statistical way of measuring the overlap between all of the different sub-tests within an intelligence test. This technique, factor analysis, assumes that an individual's score on a particular test reflects at least two components: their general level of ability, plus or minus an amount corresponding to the specific ability tapped by the sub-test. Because the general ability affects all sub-tests, its contribution can be found by looking at the correlations between all pairs of sub-tests within the overall scale. The result of a factor analysis is an estimate of the level of general intelligence, *g*, and 'pure' measures of the specific sub-components of intelligence.

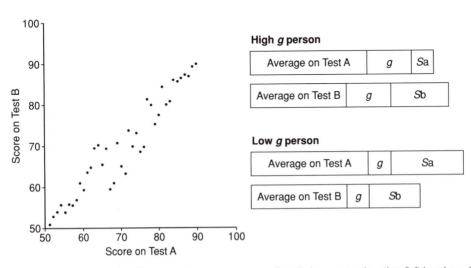

Fig. 2.   *Two tests correlate if a person's score on one predicts their score on the other (left hand graph). On the right-hand side, the bars illustrate the relative contributions of a general factor and specific factors to ability scores. A person's score on an ability test consists of the average score for that test, plus (or minus) their general intelligence (g), which is the same for all tests, plus or minus their own specific ability at that test (Sa and Sb). The Specific ability can outweigh the contribution of g, so here the Low g person scores higher on Test A than the High g person, even though they are 'less intelligent'.*

The interpretation of these results has been very controversial, and depends entirely upon the theoretical position that is taken about the structure of abilities. For those who think that ability is in essence a product of the brain's physiology, or neural efficiency, and who emphasize maturation of neural structure as the key aspect of cognitive development, *g* is most important, and the specific abilities are not really interesting. For those who think that ability is in essence a product of the individual's experience and interaction with their environment, and who emphasize the acquisition of knowledge, skills and strategies as the key aspects of cognitive development, the specific ability scores are most important, and *g* is a way of statistically eliminating the spurious influence of the testing situation. All sub-tests are, after all, tests, and although they measure different aspects of ability, they all do so in a similar way, asking a person to come up with a single correct answer for a problem as quickly as possible. In their view, *g* is simply a measure of test-answering skill, and the interesting aspects of test scores are what remain once *g* has been discarded.

Cognitive correlates

For those who argue that *g* is a real measure of ability instead of a statistical artifact, the correlation between measures of cognitive performance that are not derived from testing situations is compelling evidence of those generalizations. These cognitive correlates of *g* have been argued to include measures of choice reaction time, lexical access speed, and visual inspection time, all of which depend upon the basic speed of neural processing.

Choice reaction time (CRT) is measured by presenting people with a stimulus and asking them to classify it as a member of two or more possible sets as quickly as they can, typically by pressing a button. Jensen (1982) reported that people with high IQ had faster CRT than people with average or low IQ, although the correlation was not very large, with CRT explaining only about 10% of the variance in IQ.

Lexical access speed measures how quickly one can access the semantic meaning of a word or symbol in long-term memory (see Topic D3). This can be measured by subtracting the time taken to say whether two words or symbols are physically identical (e.g. *A* and *A* match; *A* and *B* do not) from the time taken to say whether they mean the same (e.g. *A* and *a* match; *A* and *b* do not). Hunt (1985) reported that this difference between physical matching and name

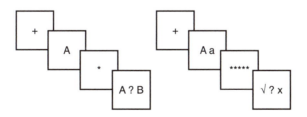

Fig. 3. Two tasks used to obtain cognitive correlates of general intelligence (g). On the left, a CRT task in which a fixation cross is followed by a stimulus, and then a 'mask' which obscures it. People then have to press one of two buttons to indicate what the stimulus was. On the right, a Lexical Access Speed task presents two stimuli followed by a mask. On 'physical match' trials, the task is to decide if the stimuli were identical, (e.g., AA or Aa). On 'semantic match' trials, the task is to decide if the stimuli are the same letter (e.g., Aa or Bb). Lexical access speed is the semantic reaction time minus the physical reaction time.

matching was a good predictor of scores on an intelligence scale, particularly a verbal scale.

Inspection time (e.g. Nettlebeck 1987) is assessed by asking people to make a CRT decision to a very briefly presented stimulus, such as two lines of unequal length, and it has been argued that people who can successfully discriminate stimuli on the basis of very brief presentations have more efficient neural processing of both the sensory information and the resulting perceptual representations. All of these measures still share a common basic methodology, however, relying on speeded responses, and so opponents of the neural efficiency argument maintain that they are simply measuring the same artifact as the pencil and paper tests.

An alternative idea is that g is related not to the speed of information processing in the brain, but to the capacity of information processing resources, and in particular, to working memory (see Topic D2). Kyllonen and Christal (1990) found that there were very high correlations between working memory capacity and reasoning ability. This suggests that being able to hold more information actively in mind allows you to solve problems faster and more accurately, and to recognize relationships between different aspects of a problem. They also noted, though, that working memory was related to speed of processing measures, so it could be argued that being able to think faster allows you to hold more information in working memory, increasing your working memory capacity. Engle (2002) has also suggested that individual differences in working memory capacity reflect differences in what he calls 'executive attention', the ability to control what you attend to and to direct your attention to specific aspects of tasks. Engle *et al.* (1999) found that working memory tasks predicted a component of g even when short-term memory tasks were taken into account, indicating that it was the control of memory rather than just memory capacity that was important. While most researchers agree that there are statistical associations between measures of intelligence, processing speed, and working memory capacity, the debate continues over the directions of causality.

Factorial models   Theorists who disagreed with Spearman's emphasis upon g have proposed models that are based on the specific factors that remain once g has been removed. These factors are usually based upon the correlations between several sub-tests within a particular domain of ability, rather than just a single test. Thurstone (1938) described seven **primary abilities**: verbal comprehension, verbal fluency, inductive reasoning, spatial visualization, number, memory and perceptual speed. In his view, it was the profile of ability across these seven dimensions that best characterized a person's aptitudes, rather than a single number, and that there were some people who were demonstrably good with words, but poor with numbers or spatial tasks, and vice versa.

Guilford (1967) took a highly compositional view of ability, arguing that all thought could be classified on a number of pragmatic dimensions reflecting the processing being carried out. The **structure of intellect** model he proposed had three major dimensions: **operations** (five basic classes of mental processes); **contents** (six kinds of representation that could be processed); and **products** (five types of responses that could be produced). He felt that in principle any operation could process any sort of content and result in any type of product, so there could be 150 different ways of measuring intellectual ability. This huge range of potential ability measures has limited the usefulness of his model, both

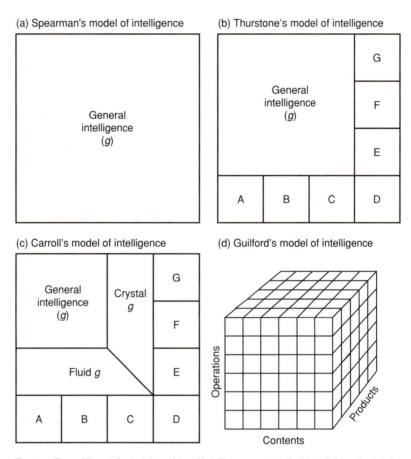

(a) Spearman's model of intelligence

General
intelligence
(g)

(b) Thurstone's model of intelligence

General
intelligence
(g)

G

F

E

A   B   C   D

(c) Carroll's model of intelligence

General
intelligence
(g)

Crystal
g

Fluid g

A   B   C   D

G

F

E

(d) Guilford's model of intelligence

Operations

Contents

Products

Fig. 4.   Four different factorial models of intelligence, each of which divides the total amount of variance in ability up in different ways. Spearman's model treats ability as unifactorial; Thurstone adds in seven specific abilities; Carroll divides the general factor into two levels. Guilford's model is completely different and divides ability into three dimensions, each with several components.

theoretically and practically, but it focused attention on the purpose for which measures were being devised. Gardner (1983) introduced the idea of **multiple intelligences** by asking what sets of ability an individual needed to use to solve the problems that were of importance to them in their own personal situations. He identified seven different types of intelligence: linguistic; logico–mathematical; spatial; musical; kinesthetic; interpersonal; and intrapersonal. While the first three of these have some similarity with Thurstone's specific factors, the others appear to some critics to be more like talents than abilities, reflecting practiced skills learnt through experience rather than a psychological ability. Gardner's view of the mind, however, is that it is **modular**, and that skills in these seven areas are governed by activity in different regions of the brain, which have different evolutionary histories (see Topic A1). As evidence for the independence of these intelligences, he points out that each area tends to have its own form of encoding or representation, a specific set of cognitive operations (see Topic E1), and a definite progression of mastery from novice to expert.

The most comprehensive attempt to relate intelligence testing to information processing measures was undertaken by Carroll (1993), who combined the results of nearly five hundred different experiments carried out by researchers across the world over 60 years. He identified three levels of ability, with the lowest level being specific to the performance of a particular **elementary cognitive task**. These specific factors were all correlated, but he argued that they were best explained by a number of general ability factors, such as lexical processing and visuo–spatial processing, and including two abstract factors previously described by Cattell (1971), **fluid** and **crystallized** ability, labeled g_f and g_c. These reflect on the one hand the ability to deal with novel problems, and to come up with creative solutions, and on the other the ability to apply learnt skills and known solutions to problems. All of these general levels of ability were also correlated, leading to the familiar measure, g.

J3 NATURE AND NURTURE

Keynotes

Genetics	There is debate over the respective contributions to intelligence of a person's genetic inheritance and the environment within which they are brought up. This has become politicized by those who believe that racial differences can be identified in measures of g. It has been difficult to resolve this debate because people with similar genes also share similar environments, and there is widespread misunderstanding of the way genetic influences are expressed. The current estimate is that half of the variation in intelligence in the population is due to genetic factors, and half to environmental factors.
Twin and adoption studies	Identical (monozygotic) twins have identical genes, and their IQs correlate as strongly as the same person tested twice. Non-identical (dizygotic) twins share only half their gene on average, and their IQs have a correlation of 0.57. The difference between the correlations is taken to indicate the importance of genetic factors. Siblings, who also on average share half their genes, but who also are brought up at different times, have IQs that have a correlation of 0.45, showing that environment also has an influence. Adoption studies have shown that the intelligence of adopted children is equally strongly correlated with both their biological and adoptive parents when they are young. By the time they are adults, only the correlation with their biological parents remains, although it is low.
Environmental effects	Deprivation leads to poor intellectual attainment, and it has been argued that enriching a child's environment can improve their attainment. Attempts to do so include the Head Start program in the US, and training of Venezuelan children on logical reasoning tasks. Worldwide measures of IQ suggest that it is rising by about three points every ten years. This Flynn effect may be due to improved nutrition, or to an increased emphasis on test-taking in education.
Race	Findings that African Americans score on average 10 to 15 points lower on IQ tests than European Americans, while Asian Americans score higher, have been interpreted as indicating racial differences in intellectual ability. The difference between people within the same racial group is still larger than the differences between groups, so knowing someone's race does not tell you how clever they are. The tests are also standardized on European Americans, and so finding that different groups have different scores is not informative. In a study that used blood tests to compare genetic markers of African descent with IQ, the only one that showed any relationship was the one that influenced skin color, suggesting that it is the way people are identified within society that is important rather than the genes themselves.

Related topics	Language development (H3)	Intelligence testing (J2)

Genetics   The question of whether variation in intelligence is primarily a result of genetic inheritance, or whether it is primarily due to environmental influences upon development, is even more controversial than the argument about the meaningfulness of g. If intelligence is largely innate, then it could be argued that there is no point providing education and social opportunities to those who simply will not be able to benefit; on the other hand, it could be argued that identifying those with limited abilities allows them to be given remedial and supportive education to allow them to overcome their limitations (indeed, this was the motivation behind the development of Binet's original tests). The argument has become even more politicized by the intervention of those who believe in racial differences, who point to evidence from the US that groups with different ethnic backgrounds have different mean scores on standardized intelligence tests. In fact, like many aspects of human physical and psychological development, intelligence is influenced by many factors, including both genetics and upbringing. The relative contribution of the two aspects is very difficult to disentangle for two reasons. Firstly, people who share genetic backgrounds also share environmental backgrounds, and the more alike they are genetically, the more likely they are to be treated identically. Secondly, there is a widespread misunderstanding of the way that genetic influences are expressed through development. Genes can lead to a range of potential developmental outcomes, but sometimes their expression only occurs in particular environmental circumstances, and the nature of their effect is then also dependent upon the environmental circumstances.

The current estimate is that environment and heredity both contribute around 50% of the variance in intelligence. Note that this does not mean that half of a person's intelligence comes from their parents, and half from their upbringing. It says nothing about an individual's intelligence, only that half of the differences within a population are attributable to environmental influences, and half are due to genetic factors. If all environmental influences could somehow be made identical, then obviously none of the variation in intelligence within the population would be attributable to environmental differences, and it would become 100% heritable. If sex were outlawed, and all babies were genetically identical clones, then genetic differences would vanish and environment would account for 100% of the variation in intelligence within the population. In the less extreme situation where a population all share some aspects of their upbringing in common, environmental factors have less effect upon the resulting variation, although they inevitably still contribute to the range of scores, making it higher or lower if they are favorable or unfavorable. The more equitable a culture is, the larger the apparent effect of genes on variation in intelligence. The more unfair it is, the lower the apparent effect. Similarly, the more genetically homogeneous a culture is, the less variation there is in genes, and the lower the contribution genes have to variation in intelligence; the less homogeneous it is, the greater the influence of genetic differences on variation in intelligence. The relative proportion of the variation accounted for by one factor or another within a population is therefore not very meaningful.

Twin and adoption studies   The strongest evidence for the contribution of genetic factors in intelligence comes from the study of identical (**monozygotic**) twins, who have developed from the same egg and sperm and so who have identical genes. The correlation between the intelligence scores of these twins is around 0.85, meaning that 60 to 70% of the variation in one twin's intelligence can be predicted by the other twin's intelligence. This is about as high a correlation as could be expected, because it is roughly the same as the correlation of scores from the same person tested twice. It compares to a correlation of 0.57 for non-identical (**dizygotic**) twins, which accounts for thirty to forty percent of variation in intelligence. Dizygotic twins develop from different eggs and so have different genes, although because they share a mother and father, they still have on average 50% of genes in common.

Proponents of the heritability of IQ point to the reduction in the correlation as the shared genetic material decreases as proof that genetic factors in some way cause differences in IQ. Opponents of heritability argue that even though all twins in these studies were brought up together, parents of monozygotic twins go to extraordinary lengths to treat them identically, far more than do the parents of dizygotic twins, and so even minor differences in environment are minimized. Furthermore, the correlation for ordinary siblings, who also have on average 50% of their genes in common, but who have been born at different times, is only 0.45 (explaining 20 percent of variation). The reduction in the correlation from dizygotic twins to sibling pairs cannot be explained genetically, but must be due to developmental differences, especially environment. All of these studies have collected data from reasonably well-off families, limiting the potential contribution of environmental differences and hence potentially inflating estimates of heritability.

Adoption studies look at the correlation between children, their biological parents, and their adoptive parents. The rationale is that if genetic inheritance is most important in determining IQ, then the largest correlation should be seen between biological parents and their children; if genes are not important, it should be no higher than the correlation between adopters and their adopted children. The Texas Adoption Project tested children twice, ten years apart. At the first test, the correlation between children and their biological parents was about the same as that between children and their adoptive parents, implying that genes and environment had similar levels of influence. By the time of the second test, however, the correlation between children and the adoptive parents had dropped away, suggesting that as they became adults only the genetic influences remained. A problem with this conclusion is that the adoptive

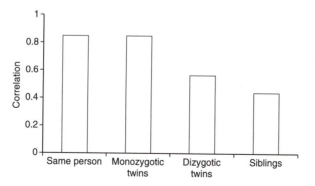

Fig. 1. *The relative contributions of genetic and environmental factors to IQ influences correlation between the scores of relatives. The scores of Monozygotic twins (who are genetically identical) correlate as closely as those of the same person tested twice. The scores of Dizygotic twins, who share 50% of genes on average, have a lower correlation, but it is higher than for siblings, who also share 50% of genes on average, but who will have had greater difference in developmental experience.*

parents did share the same environment as their children during the study, but there was no assessment of the environment in which the adoptive parents had grown up, and so there is no reason to expect a large correlation between them and their adopted children.

Environmental effects

It has long been recognized that deprivation leads to poor intellectual attainment. Early studies on children who had missed all or some of their years of schooling found that they scored less well on intelligence tests. It has also been argued that enriching the environment can therefore increase intellectual attainment. In the 1960s the Head Start program was introduced in the US in an attempt to provide preschoolers with the type of environment that cognitive developmental psychologists thought at the time would be most conducive to the acquisition of intellectual abilities. Although direct measures of intelligence were not taken, by the time they had reached high-school, children who had been on the program were further advanced academically than a matched control group, and had fewer behavioral problems. In Venezuela, the government initiated interventions to raise the mean IQ of its population, by training underprivileged school children on logical reasoning and creative thinking, resulting in increased scores on subsequent testing. It is too early to assess the long-term consequences of this project, but one problem is that through repeated assessment as part of the intervention the children are becoming skilled at tests, making assessment impossible.

A strange finding that might be attributable to environmental improvements or to increased familiarity with testing situations is that IQ seems to be increasing gradually with time, at a rate of three points every ten years. This is not small: it means that the average intelligence of children born now will be one whole standard deviation higher than that of their grandparents. This is known as the **Flynn effect**, since it was first documented by Flynn (1987), and is too rapid to be due to changes in human genetics. One possibility is that infant nutrition has been improved across the developed world (where most IQ data is collected), another is that IQ testing has had such a pervasive influence on educational policy that children are now being taught more to be good test-takers than to be well educated. In effect, many of those who obtained low scores on tests did so for environmental or developmental reasons. Remedying these deficits allows their descendants to obtain higher scores, gradually raising the mean score of the entire population.

Race

The idea that different racial groups have different mean levels of intelligence, due to their different genetic inheritance, has a history as long as the discovery of heredity. Authoritarian regimes, such as that in Nazi Germany in the 1930s, have used the argument as support for policies of **eugenics**, literally meaning 'good breeding', involving the sterilization and murder of minority groups. Recently psychological evidence has discredited the idea that there is any direct relationship between an individual's **genotype** (i.e. their genes) and their intelligence, although there may be some relationship between measures of intelligence and certain visible aspects of their **phenotype** (i.e. the expression of their genes).

A commonly cited finding is that African Americans score on average ten to fifteen points lower than European Americans, while Asian Americans score higher, an argument popularized in Murray and Herrnstein's *The Bell Curve*, whose title alludes to the shape of the normal distribution of intelligence scores in the population (see Topic J2). There are two problems with this argument. First, the differences within these ethnic groups is still far larger than the differences reported between groups, so knowing that two people belong to different groups does not tell you very much about which person will have the higher intelligence score. Second, there is little point in applying tests standardized on one population (i.e. European Americans) to another population (i.e. African Americans) and then comparing scores: most Americans of any descent would score very poorly on tests comprised of sub-scales which measured canoe navigation across the open Pacific, food collection in the Australian bush and rapid discrimination of tonal intonation patterns, yet these skills would all correlate highly with g in certain populations.

The data used to support the argument for racial differences is very weak. One study compared IQ with self-reported penis size (measured with a piece of string) and ejaculation distance, for example, arguing that genotypes with high levels of fertility did not need to be as clever as those with lower levels. They also use very superficial measures of race which do not correspond closely to actual genetic variation. In a study that did use blood tests to compare intelligence with particular genes known to be common in people of African descent or European descent, it was found that the only gene that predicted intelligence was the one that also determined skin color (Loehlin *et al.*, 1973; Scarr *et al.* 1977). In other words, no matter what other genes you have, if you grow up in America with a dark skin then you are likely to score less well on intelligence tests. Clearly, this is more likely to be due to the way you are treated in American society and by the expectations you are allowed to have for yourself than by some neurological side effect of melanin synthesis.

Because the Flynn effect operates on those who score below the mean of the population, it will eventually remove any differences in mean IQ score between subgroups in developed countries, making arguments about race and ability irrelevant. Sarason and Doris (1979) describe how Italian Americans in the early 20th century had lower than average IQ, and that this was then explained as due to their low genetic quality, their being the least fortunate group within their native society, forced by poverty to emigrate to the US. By the end of the century, Italian Americans had on average an IQ above 100. Sharing the same environment, goals and aspirations as the rest of the population, and not being placed at an institutional or social disadvantage, allowed the descendants of those immigrants to develop in the same way as the rest of the population. They did not necessarily become any cleverer than their great-grandparents, but they were able to behave in a similar way to the rest of the population when confronted with an IQ test.

K1 ISSUES IN CONSCIOUSNESS RESEARCH

Keynotes

What is consciousness?	The term 'consciousness' covers cognitive functions such as attention and voluntary action; sensory experiences or qualia; and states such as being awake or dreaming.
Do cognitive psychologists study consciousness?	Much of cognitive psychology involves studying people's responses to conscious perceptions or knowledge. Consciousness is implicated in many cognitive models, yet few of these actually attempt to explain consciousness itself.
The mind–body problem	Although mental and physical processes interact, they seem to have different properties. Dualists argue that they are different classes of thing, whereas extreme materialists argue that mental terms are misleading and only brain research will reveal the nature of consciousness. Most consciousness researchers assume that mind is a property of the brain, and can be understood by researching its neural correlates and cognitive functions.
The problem of subjectivity	Our introspections are personal; we cannot know what it is like to be someone else. We have no objective, publicly observable data on conscious experience, so consciousness researchers must infer conscious states from people's overt behavior.
The hard and easy problems	The easy problem of consciousness is explaining how modular cognitive systems share the information they generate, so that we can talk about things we perceive and use memory to help solve problems. The hard problem is explaining how neural and cognitive processes cause conscious experience.

Related topics	Cognition (A1)	States of consciousness (K2)

What is consciousness?

The term 'consciousness' covers a wide range of cognitive functions, including voluntary action, decision making, attention, awareness, and self-monitoring. It also covers sensory experiences, for example what it feels like to taste chocolate as opposed to cabbage or to look at something red rather than blue. Philosophers call these particular conscious experiences **qualia**. Being awake, or dreaming while asleep, are conscious states whereas being deeply asleep, anaesthetized or in a coma are unconscious states. Hypnosis, meditation and the effects of hallucinogenic drugs are termed altered states of consciousness.

Do cognitive psychologists study consciousness?

Marcel (1988) argues that psychologists do, and must, study consciousness. Even a simple task like pressing a button when you see a red light involves consciousness because you only respond if you are conscious of seeing the light. We need to understand consciousness to satisfy people's curiosity about their mental life and to explain their behavior. Consciousness is no more or less meaningful than other hard-to-define concepts like intelligence or personality.

However, many cognitive psychologists study conscious processes without actually referring to consciousness. Thus models of attention (Section C) aim to solve the problem of how we can be conscious of the important aspects of our environment while ignoring other aspects, yet they do not attempt to explain consciousness itself.

The mind–body problem

We know that mind and body are closely intertwined – a blow on the head can make us unconscious and imagining a large hairy spider crawling across up your arm might make you break out into a cold sweat – but they seem to have very different properties. For example, if you imagine traffic lights changing from 'stop' to 'go', your image changes from red to green. Presumably the neurons that are doing the imagining do not change color. The mind–body problem is thus the problem of how consciousness, with all its feelings and experiences, can arise from the physical brain when they seem such different types of thing.

Solutions to the mind–body problem vary. At one extreme is the Cartesian (after Descartes) or **dualist** position, which argues that mind and body are separate things. Dualists face the problem of explaining how an immaterial mind interacts with the physical body, for example, how does your desire to drink a cup of coffee cause your arm to move and pick up the cup? At the other extreme is **eliminative materialism**, which argues that mentalistic terms like desire are misleading and should be eliminated from consciousness research, which instead should focus on discovering how the brain works.

The vast majority of psychologists fall between these extremes, assuming that consciousness is a brain function and that terms like 'intention' or 'joy' are just alternative ways of referring to the neural processes underlying voluntary action or emotion (see discussion of functionalism and materialism in Topic A1). Much consciousness research aims to discover the **neural correlates of consciousness**, those brain processes that always accompany conscious experiences. Cognitive psychology helps identify the cognitive processes that accompany conscious experience, the **cognitive correlates of consciousness**, and helps reveal the functions of consciousness.

The problem of subjectivity

Scientific research requires publicly observable data, so that people can test each other's theories by running their own experiments. A problem in consciousness research is that we only have access to our own conscious experiences and we can never truly know what being conscious (or seeing red or feeling sad) is like for someone else. Nagel (1974) describes this problem as the problem of knowing what it is like to be a bat. If we know enough about echolocation and bat aerodynamics, we could imagine flying around at night listening for moths. But we would only be imagining what it would be like for us to be a bat, not what it is like for a bat to be a bat.

The reliance on 'first person' data, rather than publicly observable 'third person' data, contributed to the downfall of introspectionism (see Topic A3) and remains a problem in consciousness research. Not only is introspection not verifiable, but it can be misleading. Nisbett and Wilson (1977) asked subjects to evaluate the appearance, accent and mannerisms of a videotaped instructor. These attributes were identical in two conditions. In one condition, participants saw the instructor behaving in a warm friendly manner and rated his attributes positively. In another condition, participants saw him behaving in a cold, detached manner and rated his attributes negatively. They reported that their dislike of the attributes *caused* their dislike of the instructor, not vice versa. This study demonstrates people's lack of insight into their mental processes. We are only aware of the products of our cognitive processes, not of the processes themselves.

Despite the pitfalls of introspection, many interesting aspects of consciousness can be studied by observing people's behavior, which is presumably a reflection of their conscious state, in the same way that other aspects of cognition are studied. Conscious states can be inferred from people's behavior (including their verbal reports) in the same way that gravity is inferred from the behavior of physical objects.

The hard and easy problems

So, most psychologists and neuroscientists study consciousness by investigating the cognitive and neural processes underlying conscious perceptions, decisions, recollections, actions etc. This research addresses the problem of how we can filter and interpret an enormous amount of incoming sensory information so that we can use it to make decisions, modify our behavior, reflect on past events, plan for the future, hold conversations, recognize friends etc. Chalmers (1996) calls this the easy problem of consciousness. It is the problem of how modular neural and cognitive systems get access to information generated, by unconscious processes, in other modular systems. For example, the problem of how we are able to talk about things we remember, or remember things we talked about. This aspect of consciousness is also known as **access consciousness** (Block, 1995).

The hard problem is the problem of explaining how and why these neural and cognitive processes cause conscious experience. There seems to be nothing about neural and cognitive processes that necessitates the conscious experiences that accompany them. Knowing everything about how the brain works will not explain why certain neural or cognitive processes are accompanied by particular conscious experiences. In other words, there is an **explanatory gap** between understanding the brain and understanding conscious experience. Conscious experience is sometimes referred to as **phenomenal consciousness** (Block, 1995).

K2 STATES OF CONSCIOUSNESS

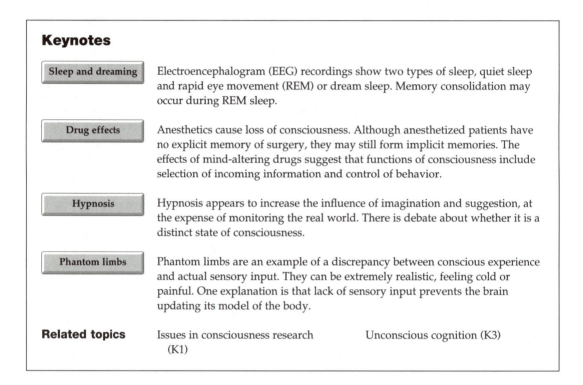

Sleep and dreaming	Electroencephalogram (EEG) recordings show two types of sleep, quiet sleep and rapid eye movement (REM) or dream sleep. Memory consolidation may occur during REM sleep.
Drug effects	Anesthetics cause loss of consciousness. Although anesthetized patients have no explicit memory of surgery, they may still form implicit memories. The effects of mind-altering drugs suggest that functions of consciousness include selection of incoming information and control of behavior.
Hypnosis	Hypnosis appears to increase the influence of imagination and suggestion, at the expense of monitoring the real world. There is debate about whether it is a distinct state of consciousness.
Phantom limbs	Phantom limbs are an example of a discrepancy between conscious experience and actual sensory input. They can be extremely realistic, feeling cold or painful. One explanation is that lack of sensory input prevents the brain updating its model of the body.
Related topics	Issues in consciousness research (K1)　　　　Unconscious cognition (K3)

Sleep and dreaming

Sleep appears to be essential to human well-being yet its functions are not well understood. Electroencephalogram (EEG) recordings show distinct stages of sleep, with three stages of quiet sleep and one stage of active or dream sleep. Active sleep is accompanied by loss of muscle tone and rapid eye movements, hence it is known as **rapid eye movement (REM) sleep**. People are much more likely to report dreaming if woken from REM sleep than from quiet or non-REM sleep. Topic D3 discussed some evidence that sleep, particularly quiet sleep, may aid memory. One explanation is that, if one stays awake after an event, then new experiences might interfere with the memory of that event. Sleeping aids memory passively by reducing this interference. A recent study by Maquet and colleagues (2000) suggests that REM sleep may help consolidate memories in a more active way. They taught participants a serial reaction time task (see Topic K3) while they were awake and used positron emission tomography (PET) imaging to reveal which brain areas were activated by the task. Subsequent PET imaging showed that the same brain areas were more active during REM sleep for the participants who had learned the task than for participants who had not attempted it.

Drug effects

Mind-altering drugs may be useful tools for consciousness research. Discovering how their effects on neural and cognitive processes are related to the altered experiences they cause may reveal something about the basis of normal conscious experience.

Drugs such as lysergic acid diethylamide (LSD) and ketamine alter conscious experience. They cause hallucinations and other perceptual disturbances such as **synesthesia**, where stimuli in one modality cause experiences in another modality (e.g. seeing purple when touching something sharp). These effects perhaps suggest that normal consciousness involves selecting incoming sensory information, restricting the amount of information that reaches higher-level cognitive processes or becomes accessible to other cognitive modules. Section C on attention discusses some models of how this selection is done. Drugs such as alcohol cause loss of inhibition, suggesting that normal consciousness involves monitoring and controlling our behavior. Models of the cognitive control of behavior suggest that we use limited-capacity processes such as working memory to select action schemas, change strategies, override habitual actions, and detect errors (see Topics C4, D2 and K5). Finn and colleagues (1999) suggest that alcohol reduces the capacity of working memory to inhibit unwanted responses.

Anesthetics are used to render patients temporarily unconscious for surgery. They produce a graded, dose-dependent loss of consciousness rather than 'switching off' consciousness in an all-or-none fashion. Although patients have no explicit memory of surgery, there is some evidence that they continue to form implicit memories during anesthesia (see Topics D1, D3 and K3).

Hypnosis

Hypnosis is a state of deep relaxation in which people are more credulous, more susceptible to suggestion, and less willing to initiate action. Under hypnosis, people seem to be able to lose their inhibitions, recall forgotten memories, and even feel no pain during minor surgery. There is debate about whether hypnosis is really an altered state of consciousness or simply a behavioral response to the social situation, the combination of relaxation and confidence in the hypnotist. Hypnosis may reduce **reality monitoring**, the normal checking of one's mental contents against the external world. Thus hypnotized people are more likely to believe the hypnotist's suggestions without checking their plausibility, are more able to maintain a hallucination (that they are 5 years old, or feeling no pain) that contradicts information from the real world (that they have a beard or are undergoing surgery), and are more likely to believe their own memory reconstructions (see Topic D4).

Phantom limbs

Experiencing a phantom limb is not strictly an altered state of consciousness, but it is an interesting example of a conscious experience that is unrelated to sensory input from the body. Phantom limbs typically occur after amputations or paralysis. They may feel warm or cold, may itch or sweat, or be painful. They are too vivid to be considered mere recollections of the removed limb, thus people may try to stand on phantom feet or lift cups with phantom hands.

Melzack (1992) proposed that phantom limbs originate in the brain. He hypothesized that the brain actively generates models of the body, a 'neurosignature' against which the incoming information can be compared to check the body is intact. If a limb has been lost, there is no sensory information with which to update the neurosignature, so it remains as it was just before the limb was lost. Ramachandran and Rogers-Ramachandran (1996) showed that 'fake' sensory information can help ease phantom limb pain. They used a mirror to superimpose an image of the remaining limb onto the felt location of the phantom limb. Moving the remaining limb thus gave the illusion of both limbs moving. After some practice with this technique, one patient reported complete 'amputation' of a phantom arm. Melzack's hypothesis suggests that our normal conscious experience is as much a top-down cognitive process as it is the result of bottom-up sensory processes (see 'Scene perception' in Topic B5 for another example).

K3 UNCONSCIOUS COGNITION

Keynotes

Perception without awareness	Patients with blindsight have no experience of seeing stimuli in the blind part of their visual field, yet can respond to those stimuli with some accuracy. Patients with visual neglect can show emotional responses to stimuli that they appear not to notice.
Unconscious processes in language	Much of language processing occurs without conscious control. Even when interpreting complex or ambiguous statements, we are often unaware of using memory to draw inferences about meaning.
Implicit learning and memory	Evidence for learning during anesthesia suggests that implicit memories can be formed in the absence of consciousness. Evidence that participants' behavior can be influenced by subliminally presented stimuli or hidden rules also supports claims of implicit learning.
Action without awareness	Many complex but well practiced tasks can be performed without conscious awareness or conscious control. Even when we feel that we deliberately perform an action, our 'free will' may be illusory.

Related topics	Automaticity (C4)	The function of consciousness (K4)
	States of consciousness (K2)	

Much cognitive psychology research into consciousness focuses on unconscious, or implicit, cognition. Finding out what the brain can do without consciousness helps us infer the properties and functions of consciousness. The topics discussed below show that consciousness is not an essential component of many cognitive processes.

Perception without awareness

Damage to primary visual cortex causes blindness in part of the visual field. Sometimes people with this condition have **blindsight**. D.B. (reported by Weiskrantz, 1986) could respond fairly accurately when asked to guess whether a stimulus was an X or an O, or to reach towards its location. But he had no experience of seeing the stimulus and was sure he was just guessing. Blindsight suggests that sensory information can be processed without us being aware of it. However, 'seeing' in blindsight is very different from normal seeing. Although above chance, people are less accurate at identifying and localizing visual stimuli in the blind field, and they do not initiate responses to them. For example, someone with blindsight who is thirsty will not reach out for a glass of water in their blind field, even though they can point to it if encouraged to do so.

Injury to the right hemisphere of the brain can cause unilateral **neglect**, a condition in which people ignore stimuli on the left-hand side of space (see Topic C2). They may leave food on the left-hand side of the plate, bump into doorways, or only half dress themselves. Marshall and Halligan (1988) showed two pictures of a house to a patient, P.S., with neglect. One house had flames emerging from its left-hand side. P.S. preferred the other house but perceived both pictures as identical and could not explain her preference. As in blindsight, it appears that some information is processed but that processing does not result in a conscious experience. It is debatable whether these findings reflect damage to single systems, so that processing occurs but does so below the threshold needed for conscious awareness, or whether there is damage to 'consciousness modules' that are separate from the lower-level systems that enable reasonably accurate 'guesses' about stimuli.

Unconscious processes in language

Many cognitive processes are unconscious, in the sense that we are unaware of their occurrence and of their products. Language provides some good examples (see Topics H4 and I2). Speech segmentation and syntactic analysis proceed without conscious control or awareness. Even understanding the meaning of sentences relies on unconscious inferences. We are not aware of consulting memory to comprehend ambiguous sentences; often this process happens so rapidly that we do not realize the sentence was ambiguous. For example, if we read that Bob was hungry so he reached for the menu, we assume that he was going to order some food rather than eat the menu.

Implicit learning and memory

Studies of unconscious or implicit memory (see Topic D3) show that we remember more than we are aware of. Implicit memory tests can be used to test **implicit learning**, that is, learning that occurs without awareness of the information being learned. Deeprose and colleagues (2004) played words to patients during surgery with general anesthesia. On recovery, patients listened to word stems and responded with the first completion that came to mind. They completed the stems with the target words (i.e. the words on the experimenters' list) more often if those words were presented during surgery than if they were not presented, suggesting that they had formed implicit memories of the words even though they had no explicit memory for surgery or the words themselves.

Implicit learning can be demonstrated using **subliminal presentation** of stimuli, that is, by presenting visual stimuli so briefly that participants are unaware of seeing them. Kunst-Wilson and Zajonc (1980) presented geometric patterns once for very brief periods and then again for longer periods. On the second presentation, each pattern was paired with another, similar pattern that had not been shown before. Participants were asked i) which shape in each pair they had seen before, and ii) which shape they liked better. They performed at chance (random guessing) on the first test but tended to choose the previously presented patterns on the second. This is an example of the **mere exposure effect**, the tendency to prefer previously encountered stimuli even in the absence of explicit memory for them. Mandler, Nakamura and van Zandt (1987) replicated and extended this finding. In some conditions, they asked participants to choose which pattern was brighter, or which pattern was darker. Participants showed a tendency to choose the previously presented patterns on both versions of the test, suggesting that unconscious processing of the patterns caused a response bias rather than a genuine change in emotion or perception.

Marcel (1983) used subliminal presentation to demonstrate **semantic priming** (see Topic D3). A 'prime' word was presented very briefly, followed by a 'mask', a pattern that prevented any lingering iconic memory (see Topic D2) of the word. Then a second word was presented and participants judged whether this 'target' word was a real word or not. Response times were faster on this task if the prime and target were semantically related (e.g. bread–butter).

An alternative to subliminal presentation is to present clearly visible or audible stimuli but hide the information that you want participants to learn. In **artificial grammar learning** experiments (Reber, 1967), participants learn strings of stimuli made according to a fixed set of rules or grammar (*Fig. 1*). They are then asked to select the 'grammatical' strings from similar strings that do not follow the rules. People typically score above chance on this task yet cannot explain the rules that led them to make their responses. In Nissen and Bullemer's (1987) **serial reaction time task**, there are four stimuli that can light up on a computer screen, and four corresponding buttons on a keypad. The participant's task is to respond as quickly as possible to the lit stimulus by pressing the appropriate key. Unknown to participants, the stimuli light up in a fixed sequence. If the sequence is changed, participants respond more slowly, suggesting they have learned something about the original sequence. Despite this, they are typically unable to say what the sequence was.

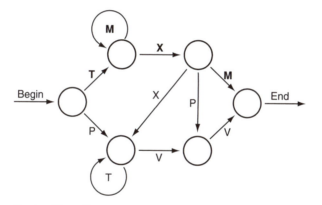

Fig. 1.   The artificial grammar used by Reber (1967), with a grammatical sequence (TMXM) highlighted. The curved arrows show letters that can be repeated, thus TMMMXM is grammatical.

One explanation of these findings is that people have acquired unconscious knowledge of the grammar or sequence. In other words, performance on these tasks demonstrates implicit learning. Critics of this view argue that people have some fragmentary conscious knowledge of the grammar, for example they may know that two stimuli tend to be paired. People do not report this fragmentary knowledge because the tests of explicit knowledge do not ask about it. However, it may be sufficient to explain their above-chance performance on the tests.

Action without awareness

We often perform well-practiced tasks, for example driving a car, without being fully aware of what we are doing (see Topic C4). Conscious control seems to be necessary only for novel or difficult tasks. Wegner and Wheatley (1999) go further and argue that, even with apparently voluntary actions, the feeling that we act deliberately through conscious **free will** is illusory. They asked people to move a computer mouse to point at an object on the screen a few seconds after they had heard the name of the object read out to them. The mouse was shared between the real participants and an experimenter pretending to be another participant. On some trials the experimenter made the mouse stop at the object instead of waiting for the participant to do so. People tended to believe they had intentionally caused the mouse to stop, even though in fact the experimenter had stopped it. This tendency increased if the word had occurred between 1 and 5 seconds before the mouse stopped, and was lowest if the word occurred 1 second after, or 30 seconds before, the mouse stopped.

K4 THE FUNCTION OF CONSCIOUSNESS

Keynotes

Function of consciousness	Conscious processes are slow compared with unconscious processes, but they enable us to behave adaptively, to change habitual responses and behave appropriately in new situations.
Could there be zombies?	Some philosophers argue that the conceivability of zombies, beings with identical cognitive and neural processes to humans but no conscious experience, suggests that conscious experience has no function, that it is an epiphenomenon. There may be no zombies, but could there in the future be computers that function just like humans but have no conscious experience?

Related topics	Cognition (A1)	Cognitive theories of consciousness
	Unconscious cognition (K3)	(K5)

The function of consciousness

Unconscious cognitive processes are fast and efficient but inflexible. Consciousness helps us avoid acting solely through habit. It lets us make novel responses to familiar situations and to use prior knowledge to respond appropriately to novel situations. However, controlled cognitive processes tend to be limited in capacity, hence conscious control of behavior is slower than automatic responding (see Topic C4). Patients with blindsight and neglect (see Topic K3) show us that although we may process information without awareness, this information does not usefully guide our voluntary behavior unless we are aware of it. Importantly for human culture, we cannot talk about topics without some conscious knowledge of them. Mind-altering drugs suggest that our normal conscious experience is the result of selecting information from the huge array of incoming information that bombards our senses. People's uninhibited behavior under the influence of hypnosis or alcohol suggests that monitoring and control of behavior are important functions of normal consciousness (see Topic K2).

Two studies illustrate the importance of consciousness for adaptive behavior. Murphy and Zajonc (1993) showed that consciousness helps us to make rational decisions rather than relying on emotional responses. They asked participants to decide (or rather to guess) whether Chinese ideograms represented good or bad concepts. Each ideogram was preceded by a happy or angry face. When the faces were presented subliminally, they biased participants' decisions about the ideograms, that is they 'primed' their performance. The faces had no effect when presented for long enough to be perceived, thus awareness of the 'primes' allowed participants to counteract the emotional biases in their performance (*Fig. 1*).

Baddeley and Wilson (1994) argued that explicit memory of our mistakes helps us to learn from them. Without explicit memory, we are prone to repeat our mis-

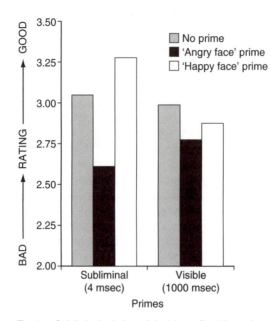

Fig. 1. Subliminal priming of decisions. Decisions about the meaning of Chinese characters were biased by the expressions in faces presented just before them, but only when presentation of the faces was too brief to be consciously perceived (from Murphy and Zajonc, 1993. Affect, cognition and awareness. J. Personality Social Psychol. Copyright © 1993 by the American Psychological Association. Reprinted with permission.).

takes indefinitely. They demonstrated this by asking people with amnesia (see Topic D4) and controls to learn lists of words under two conditions: in the 'errorful' condition, the experimenter said 'I am thinking of a 5-letter word beginning with QU. Can you guess what it might be?' The participant guessed several words, then the experimenter said 'No, good guesses but the word is QUOTE please write that down'. For the 'errorless' condition, the experimenter said 'I am thinking of a 5-letter word beginning with QU and the word is QUOTE please write that down'. For the test phase, participants were given the first two letters of each word and asked to supply the correct word. Amnesic participants learned considerably more if they had been prevented from making errors at the start (i.e. the 'errorless' condition, *Fig. 2*). The errorless learning technique has been used to teach useful skills and information to people with memory problems.

Could there be zombies?

Being able to monitor and control our behavior are consequences of being able to share information across cognitive modules. For example, being able to use memory of previous mistakes to change our behavior, or using conscious perception of an object to describe it to someone. In other words, monitoring and control are functions of access consciousness (see Topic K1). Does phenomenal consciousness or conscious experience also serve a function or is it just an **epiphenomenon** of our neural and cognitive systems (i.e. a by-product that has no causal effect in the system)? Philosophers use the concept of **zombies** to help discuss this issue. Zombies are hypothetical (imaginary) beings that have the same bodies as us and the same brains. They have the same cognitive processes as we do and consequently the same knowledge. They behave in exactly the same way as we do because they process information in the same way,

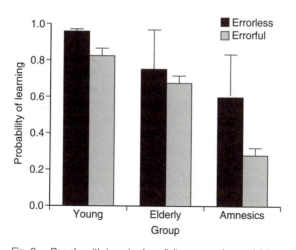

Fig. 2. People with impaired explicit memory (amnesia) learn better if they are prevented from making errors during learning. The benefit of errorless learning is much less for young and elderly controls with normal memory function. (Reprinted from Baddeley, A. D. & Wilson, B. A. (1994) When implicit learning fails: Amnesia and the problem of error elimination. Neuropsychologia, 32, 53–68. © 1994, with permission from Elsevier Science).

interpret it using the same knowledge, and use it to influence their actions in the same way. The only difference between zombies and humans is that zombies have no conscious experience.

Some philosophers argue that the conceivability of the concept of zombies suggests that conscious experience may be unnecessary. The fact that we can imagine a zombie (i.e. that the idea is coherent, even though we know zombies do not exist) suggests that conscious experience is not an inevitable consequence of cognitive and neural processing, leading some to argue that it does not have a function. This issue is hard to resolve because of the problems of subjectivity etc. in studying conscious experience (see Topic K1).

A related issue arises in **artificial intelligence** (see Topic A4). If a computer is programmed with identical cognitive processes to a human, will it be conscious? If conscious experience has no function, no effect on our behavior, then we will not be able to judge if a computer is conscious simply by observing its behavior.

K5 COGNITIVE THEORIES OF CONSCIOUSNESS

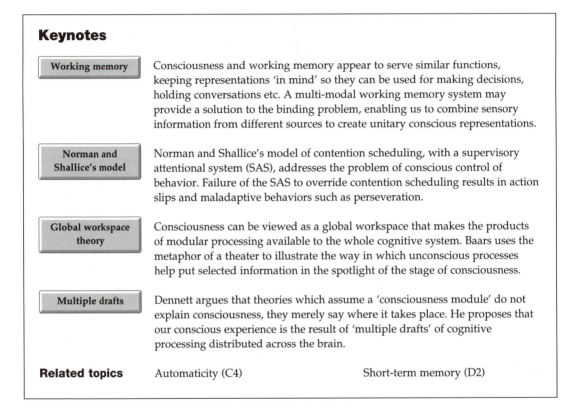

Keynotes

Working memory

Consciousness and working memory appear to serve similar functions, keeping representations 'in mind' so they can be used for making decisions, holding conversations etc. A multi-modal working memory system may provide a solution to the binding problem, enabling us to combine sensory information from different sources to create unitary conscious representations.

Norman and Shallice's model

Norman and Shallice's model of contention scheduling, with a supervisory attentional system (SAS), addresses the problem of conscious control of behavior. Failure of the SAS to override contention scheduling results in action slips and maladaptive behaviors such as perseveration.

Global workspace theory

Consciousness can be viewed as a global workspace that makes the products of modular processing available to the whole cognitive system. Baars uses the metaphor of a theater to illustrate the way in which unconscious processes help put selected information in the spotlight of the stage of consciousness.

Multiple drafts

Dennett argues that theories which assume a 'consciousness module' do not explain consciousness, they merely say where it takes place. He proposes that our conscious experience is the result of 'multiple drafts' of cognitive processing distributed across the brain.

Related topics   Automaticity (C4)   Short-term memory (D2)

Working memory

Several authors, from James (1890) to Baddeley (1993), have identified consciousness with working memory (see Topic D2), arguing that we are conscious of the contents of short-term memory and that working memory serves to retain consciousness of recently encountered or retrieved information. Although this assumption has not been directly tested, it is supported by evidence that working memory is important for conscious cognitive activities such as mental imagery (see Topic E2) and problem solving (see Topic F2). Information stored in the phonological loop and visuo–spatial sketchpad can be combined by the central executive to create multimodal representations that may form the basis of our conscious experience. Working memory may thus provide a high-level solution to the **binding problem**, the problem of integrating information from different sources to create a unified conscious percept (see Topic B4). Baddeley (2000) argued that these multimodal representations are stored in an **episodic buffer**, although his revised model does not specify whether information is conscious by virtue of being in the episodic buffer or whether it is only conscious when acted upon by the central executive (see Topic D2).

Norman and Shallice's model

Norman and Shallice (1986) tackled the problem of voluntary action (see Topic C4). They argued that often an appropriate action schema is selected from competing alternatives by **contention scheduling**, a relatively automatic process in which the currently most active option inhibits competing options and so 'wins' the competition (i.e. that action is performed). Habitual actions are more easily activated than novel actions, hence the model also contains a control mechanism for overriding habits, detecting errors, and deliberately deciding a course of action. This conscious control component is called the **supervisory attentional system** (SAS, also known as the supervisory activating system). It oversees the contention scheduling process, increasing the activation of alternative action schemas that would otherwise be too weak to be selected. **Action slips** (e.g. adding milk when intending to make black coffee) are interpreted as failures of the SAS to override habitual action schemas. Behavioral problems such as **perseveration** (persisting with an earlier response when circumstances require a new response), often seen following frontal lobe damage, can also be explained as failures of conscious control of action.

Cooper and Shallice (2000) implemented the SAS model in a computational model. When lesioned, the model simulates the pattern of behavior seen in patients with **action disorganization syndrome**, including frequent errors in goal-directed action and **utilization behavior** (use of objects that are inappropriate to the task, simply because they are close to hand).

Global workspace theory

Baars (1988) views consciousness as a 'global workspace', a means of making the products of cognitive modules available to the whole cognitive system. He uses the analogy of a theater. All the actors on the stage are potentially visible, but we only see the actor currently in the spotlight. Working memory forms the stage of consciousness, the contents of working memory being actors on the stage. The contents of working memory are potentially conscious but do not actually become conscious until selected by the central executive (the stage director). The actors are supported by unseen stagehands, the cognitive processes of edge detection, language processing, memory retrieval and so on. Once in the spotlight, an actor becomes public property, that is, conscious representations are available to other cognitive processes.

Multiple drafts

Dennett (1991) has criticized approaches that associate consciousness with the operation of a high-level cognitive module such as the central executive or supervisory attentional system. Dennett argues that identifying a particular brain region or functional system as the seat of consciousness does nothing to explain consciousness. He refers to such sites of consciousness as **Cartesian theaters**, a reference to Descartes' unsubstantiated claim that mind and body interacted in the pineal gland. Cartesian theater approaches lack **cognitive economy**, because they assume that representations in one cognitive module are passed onto, and re-represented in, the consciousness module. Rather, Dennett argues, information only needs to be processed once. The contribution of this information to conscious experience varies according to which other cognitive processes are operating at the same time. There is not a single stream of processing that culminates in a particular representation becoming conscious. Rather, there are parallel streams of cognitive processes. The products of these processes represent 'multiple drafts' of our interpretation of the world. Which draft we become aware of depends on how we interrogate the system to meet current task demands. Dennett uses the analogy of a manuscript that is still being revised. The different authors, editors and reviewers each possess a slightly different version of the manuscript. The question 'what is in the manuscript?' will produce different responses depending on who we ask and when we ask them.

Dennett argues that our apparently unitary and coherent stream of consciousness does not reflect a unitary stream of information processing. Rather, when we ask 'what have I just been doing?', we become conscious of our memories of recent events and fill in the gaps between them. For other examples of our ability to 'fill in', see the discussion of change blindness and inattentional blindness in Topic B5.

L1 THE PRIMACY OF AFFECT

Keynotes

James–Lange hypothesis	The commonsense view is that we perceive an event, which causes an emotional state, and there are then bodily changes. James and Lange argued that the bodily changes directly followed the perception of the event, and that it was the perception of these changes that caused emotion. This put affect before cognition in the perception of emotion. They were criticized by Cannon and Bard because the range of bodily changes was smaller than the number of emotions, and changing or artificially stimulating the nervous system did not cause emotion.
Two factor theory	Schachter argued that emotion required both physiological changes and an event that the changes could be attributed to. If either occurred alone, no emotion would be felt. The exact emotion felt depended upon the cause that the changes were attributed to, and hence the appraisal of emotional situations was crucial in determining emotional perceptions. This theory had most influence in social psychology, where the effect of social situations upon emotion was studied.
Zajonc–Lazarus debate	Zajonc suggested that emotion and cognition were two separate systems, acting in parallel. Emotional perception dealt with liking (or not liking) a stimulus (the **preferenda**), cognitive perception with identifying it (the **discriminanda**). Lazarus debated with him, arguing that cognitive processing was cumulative and did not need to produce a complete output before emotional responses could occur. The debate foundered upon incompatible definitions of cognition.
Facial feedback	Our facial muscles respond rapidly to emotional stimuli and are used to signal our emotional state to others. They may form a more informative basis for the perception of affect than visceral states. Manipulating facial muscles into positions that mimic emotional expressions does seem to induce the corresponding affective state. The evidence seems to indicate both that cognitive activity can induce bodily changes, and that bodily changes can induce cognitive states of emotion: the two interact.

Related topics	Sensation (B1)	The function of consciousness (K4)
	Unconscious cognition (K3)	Mood, memory and attention (L2)

James–Lange hypothesis	Emotion is a topic that is studied by many branches of psychology, because it is associated with particular physiological changes, and with activation of distinct regions of the brain. From a cognitive viewpoint, it interacts with our memory, perception, decision making and social behavior. The involvement of the body, and the neurological and physiological aspects of emotion, make it a challenging area for cognitive theory. Much of the research has focused on the question

of whether our emotions cause these physical '**affective**' changes, or whether the perception of an emotion follows from them.

The commonsense view is that we have an emotion, such as happiness or anger or fear, and because of this essentially mental state, our body reacts appropriately. We then 'feel' the euphoria or arousal, in the form of raised heartbeat, clammy palms, butterflies in the stomach, and so on. In the late nineteenth century, James and Lange turned these notions on their heads, by arguing that the sequence of events was the other way around: that we reacted physically to events in the world or the mind first, and that we only perceived an emotion when we became aware of these changes in our body. James supported this argument by pointing out that if you just imagine or recall a situation without any **physiological reactions**, then there is no sense of emotion. The thought seems merely intellectual, cold, and flat. It is without **affect**. It is only when bodily changes occur too, that the thought becomes emotional. Just thinking about an emotion is not enough to have the emotion.

This argument was of course controversial. Physiologists such as Cannon and Bard pointed out that the nervous system by which we perceived changes in the body was very limited, and that many different emotions involved similar patterns of physiological arousal. Animals with damage to their visceral nervous system continued to display 'emotional' behavior. Visceral and physiological changes were also slow, unlike the rapid onset of emotion, and lasted longer than the emotion. A doctor called Marañon gave volunteers adrenalin (epinephrine), which causes the same bodily arousal as fear, and reported that although they correctly reported the physiological changes, they did not report feeling emotional. All of this evidence undermined James and Lange's theory about the primacy of affective change in the body for the perception of emotion, and it was generally accepted that mental activity came before changes in body state. In the middle of the twentieth century, the rise of behaviorism seemed to confirm this, with the physiological changes being seen as biologically appropriate learnt responses to stimuli that had come to be associated with emotional situations. Emotions were behavioral responses, just like any other; they could be enhanced or suppressed, but their occurrence was a matter of learning.

Two factor theory

It was only with the rise of cognitive psychology in the last third of the century that it became acceptable to study mental states such as emotions once again. Interestingly, the first cognitive theories of emotion appeared to be a refinement to the James–Lange arguments. In a classic study, Schachter and Singer (1962) gave volunteers an injection of either a placebo or an arousing drug but only informed some of the drug group correctly about the effects it would have. They then put volunteers in either a 'euphoric' condition or an 'anger' condition. In each condition, they were asked to go to a room to fill out a questionnaire. In the room was a 'stooge', an experimenter, play-acting the role of another volunteer. The stooge behaved in an overtly emotional manner, either happily playing around or angrily ripping up the questionnaire and storming out of the room.

Schachter and Singer argued that two factors, physiological change and apparent cause, were both necessary for an emotional perception to occur. If either occurred alone, there would be no emotional change. Only when there was both a physiological change and something to attribute that change to, would emotion be felt, and the emotion that resulted would depend upon what it had been attributed to. This meant that the same physiological changes could lead to different emotions. In their experiment, the drug-informed groups, who felt physiological changes but knew they were due to the drug, should not have felt much emotional reaction. Those who had the placebo, and so who had no change in arousal, should also not experience any emotional change. The drug-uninformed groups, who had the same physiological changes, but did not realize that they were due to the injection, should report emotional reactions mirroring the behavior of the stooge (*Fig. 1*). This **attributional theory** said that physiological changes preceded the perception of emotion, like James and Lange, but unlike them it said that cognitive activity could influence the emotion that was felt.

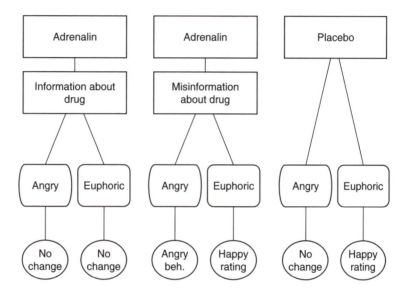

Fig. 1.   In Schachter & Singer's (1962) test of the Two Factor Theory of Emotion, volunteers who were given an arousing drug, but who were not correctly informed about its effects, showed a tendency to feel emotions consistent with the social situation that they were in. Those who were correctly told about the effects of the drug correctly attributed their arousal to the drug, and did not experience a change in emotion. Those who were placed in similar situations but did not receive a drug did also experience some emotional reactions, perhaps because the situation itself was arousing.

The results did not match the predictions of the two factor theory entirely, but the general pattern was in the right direction. In the euphoria condition, the drug-uninformed group did report feeling happier than the drug-informed group. In the anger condition, none of the three groups differed significantly in their reported feelings, perhaps due to the social constraints upon reporting angry feelings. On an objective measure of 'angry behavioral acts', though, the drug-uninformed group made more angry acts than the other two groups.

Before Schachter and Singer's work, emotion reflected biologically deter-mined, learnt responses. Social and situational factors could influence the expression of emotion, but not the emotion itself. Afterwards, emotion was a consequence of a cognitive process, and social and situational factors could determine the nature of the emotion felt, given some affective changes in the body. For this reason, this work had most influence in social psychology, where much research was conducted into how people appraised situations to determine the appropriate emotion. Cognitive psychology neglected this new emphasis upon affective reactions in the body until 1980, when Zajonc used his presidential address to the American Psychological Association to criticize what he saw as an overly sequential, information processing view of emotion.

Zajonc–Lazarus debate

Zajonc typified cognitive theorists as arguing that emotion could only occur at the end of a chain of perceptual and semantic processing that allowed one to recognize something that was happening, and to retrieve enough information from long-term memory to decide whether to be happy or sad, angry or frightened. A rabbit that reasoned about its emotions like this, he argued, would still be pondering the rustle in the grass by the time the snake had pounced upon it. The whole point of emotion was to alert us to danger or to opportunity and to focus our cognitive processing upon it. He argued that emotion and cognition acted in parallel, as two independent systems. While the conventional cognitive sequence of processing dealt with the attributes of a stimulus that allowed us to identify it and discriminate it from others (the **discriminanda**), an emotional processing route dealt with the attributes that allowed us to decide its affective nature and whether we liked it or not (the **preferenda**).

Lazarus joined the debate by responding that Zajonc's description of cognitive models was overly sequential. He saw cognition as starting as soon as a stimulus occurred, so that very salient stimuli could cause some degree of emotion very rapidly, before they had been completely recognized or identified. The longer they were processed, the more they could be recognized, and the stronger the emotion could become. He cited evidence from some studies using very brief (**subliminal**) presentations of nonsense syllables, which in a training phase had been paired with mild electric shocks. In the later testing phase, his volunteers could not report the syllables, but they displayed physiological responses to them. This, he argued, showed that sufficient processing was occurring to support the affective response (and hence provide the basis for an emotional appraisal) even though it was insufficient to support identification. Even the rabbit who runs from a rustle in the grass has some knowledge that rustling grass is dangerous, and so it starts to run, even before it has completely identified the source of the rustling noise.

Evidence that emotional responses occur early in the processing of a stimulus come from demonstrations of the **mere exposure effect** (presenting stimuli subliminally, so briefly that they cannot be consciously recognized, leads to participants to like them more than unpresented stimuli, when they are given a subsequent 'preference' test; see Topic K3), from **affective priming** (subliminal presentations of emotive stimuli affect judgments of visible stimuli; see Topic K4), and from **unilateral neglect** (where people with brain injuries that prevent them from consciously perceiving stimuli on one side of their visual field still experience emotional responses to those stimuli; see Topic K3).

Facial feedback

The debate between Zajonc and Lazarus continued for several years, because the evidence each cited could be used to support the other's position. They held incompatible definitions of what cognition was: for Lazarus, any response showed that some degree of cognition had occurred, even if it was not complete; for Zajonc, responses to sensory input were not cognitive; some refinement or completion had to occur. The debate spurred Zajonc to carry out further studies into the affective components of emotion, particularly into the role of **facial feedback**. The facial muscles are used extensively in emotional reactions, and they signal our emotional state to others. Charles Darwin had pointed to the facial expressions of babies as evidence that emotions were inherited, not learnt (and to the expressions of animals to show that they too had emotions, and that we might therefore share an evolutionary history with them). The perception of changes in facial muscles could provide a fine-grained and rapid source of physiological information as input to cognition.

The evidence for facial feedback is intriguing. For example, stories containing vowel sounds that make people pout when they read them are rated as less likeable than stories without these vowel sounds. Holding a pen in your mouth crosswise, without touching it with your lips, creates a similar muscular pattern as smiling, and induces a more positive mood than holding it pointing outwards with your lips closed around it. Taken together with the evidence from the two factor theory, it shows that we can be 'fooled' into feeling emotional changes by manipulations in the state of our body. However, it contains some problems. If our facial muscles signal our emotional state to others, then that state ought to exist before (or at the same time as) the muscular change, and not be caused by them. The resolution of the debate about whether affect is primary, or whether cognition is primary, may come through a recognition that both influence each other: being in a cognitive emotional state has bodily effects; bodily changes can be detected and can cause a cognitive emotional state.

L2 MOOD, MEMORY AND ATTENTION

Keynotes

Beck and Bower

In clinical psychology, Beck's theories about the cause of emotional disorders such as anxiety and depression signaled a break with behaviorism. He felt that disorders were caused by dysfunctional cognitions, thoughts and memories. Bower placed this approach in the context of a spreading activation model of memory, with his associative network model (ANM). Emotions were represented within memory as nodes that could be activated just like any other concept. Depressed people had negative emotional nodes with overdeveloped links, so that they were activated more easily; similarly, anxious people had an overlinked threat or danger node.

Mood congruency

Theories of emotion that rely on memory activation predict that items that are congruent with one's mood should be easier to learn (mood-congruent learning) and easier to recall (mood-congruent recall). This is because the nodes representing those items will share activation with the mood node. While this does seem to be true for depression, it is not the case for anxiety (although it may be true for implicit memories). Thus some moods may be related to a memory bias, but other moods may have other cognitive causes.

Attentional bias

Cognitive tasks have shown that anxious people, and normal people who have been made anxious, show an attentional bias towards material related to their anxiety. In the emotional Stroop task, people take longer to name the color that threatening words are printed in, compared with neutral words. In the dot probe task, attention is drawn towards the location of a threatening word on a computer screen, and stimuli that replace it (such as small dots) can be detected faster than dots that replace neutral words.

Related topics

Divided attention and dual task performance (C3)	The primacy of affect (L1)
Short-term memory (D2)	Emotion, perception and performance (L3)
Representations (E1)	

Beck and Bower

Clinical psychologists have motivated much of the theoretical and empirical progress regarding the relationship between mood and memory, because of the importance to them of the development of effective treatments for the 'affective disorders' of **depression** and **anxiety**. Behaviorist models of learning had seen these illnesses as maladapted patterns of response to stimuli, curable by relearning, by extinguishing, or by shock therapy; treatments that had met with limited success. Beck argued that depression and anxiety were caused by dysfunctional cognitions, thoughts and memories, and that addressing these in therapy could resolve the disorders. Depression, for example, was characterized by automatic negative thoughts about the self, about the world, and about possible future events. Anxiety was characterized by overactive schema for danger and threat, which take control in response to any stress.

Although successful in turning the attention of psychologists towards the cognitive aspects of mood and memory, Beck's ideas were criticized because they did not explain how the negative or danger schemata developed initially, and the theory relied on the contents of cognition rather than how the contents were processed. It was thus difficult to test, and could not be related to developments in cognitive theory. Bower was one of the first theorists to propose a more formal model for the role of mood in memory, which could explain why some people developed dysfunctional mood states. His **associative network model** (ANM; Bower, 1981) was based on a model of memory called **spreading activation**, in which memories are stored as patterns of activation between concept units or nodes (see Topic D3). Related memories trigger each other because they share links to the same basic nodes. Bower suggested that certain basic emotions could also be nodes in memory. Events such as bereavement, social embarrassment, and failure to achieve goals would all be linked to a 'depression node', while events like physical threats, risky situations, and forthcoming exams would be linked to an 'anxiety node'. Just thinking about these events would be sufficient to trigger the appropriate mood. Nodes could become overdeveloped when thoughts about a non-emotional event triggered an emotional memory, which in turn triggered the emotion node. A link would then be formed between the non-emotional event and the emotion node, and if the emotion led to a change in behavior, this could become self-fulfilling, in that the emotion might actually bring about the depressing or anxiety-provoking outcome.

Mood congruency

Bower's ANM was a model of normal as well as dysfunctional emotional thought. It led to some hypotheses about emotional thought: mood-congruent learning, mood-congruent retrieval, and **thought congruity**. Mood-congruent learning occurs because the emotional node that is active when material is presented raises the activation of other nodes that are linked to it; thus a stimulus that is also congruent to the mood and which uses these nodes will be recognized and learnt better than one that is incongruent (and so whose component nodes are not activated). Similarly, mood-congruent retrieval occurs because nodes linked to the currently active emotional node have a raised level of activation, and so memories that use them are easier to recall than memories that do not (see Topic D3). Thought congruity is a combination of these two processes, by which an individual's stream of free associations, thoughts and perceptions of ambiguous or non-emotional stimuli will show congruence with their current mood (and will be different to the thoughts that occur when they are in a different mood).

Mood-congruent retrieval has been shown in both clinical and non-clinical samples. Lloyd and Lishman (1975) asked depressed patients to recall either happy or sad memories, prompted by neutral words. The patients could recall sad memories faster than happy memories, and the size of the difference was related to the severity of their depression. Teasdale and his colleagues manipulated the mood of non-clinical volunteers by asking them to read and think about a list of self-related negative statements (the **Velten procedure**). The speed of recall and amount recalled of negative memories again increased with increasingly negative mood. Mood-congruent learning was demonstrated by Bower *et al.* (1981) when they asked volunteers to listen to a story about a happy and a sad tennis player. Volunteers who had been manipulated into a sad mood learnt 80% of the facts about the sad player, but only 45% of the facts about the happy player. There was no such bias when mood was manipulated at retrieval, showing that the effect was due to learning rather than retrieval. Matt *et al.* (1992) presented different groups of people with lists containing positive and negative words, and found that students who are not depressed recalled 9% more positive words; students who had been manipulated into a temporarily depressed state recalled both sets equally; and clinically depressed patients recalled 10% more negative words.

There is some evidence that the effects of mood may apply to explicit memory (memories that we know we have) rather than to implicit memory (memories that we cannot report, but which influence our behavior). Denny and Hunt (1992) showed mood-congruent memory biases in depressed patients for word lists on an explicit test (where they were asked to recall as many words as they could from the list), but not on an implicit test (where they were asked to complete as many word fragments as they could, some of the fragments coming from the list that they had learnt).

Bower's ANM predicts that these effects should apply equally to anxious people as well as depressed people. While the evidence for a negative bias in depression is strong, there is little evidence for such a bias in anxiety. Watts *et al.* (1986) could not find any evidence of biased recall for spider-related material in people with spider phobia. Mathews *et al.* (1990) compared explicit and implicit memory tests for threatening and non-threatening material in anxious patients, and found the opposite pattern of results to those that had been found in depressed patients: no bias on the explicit free recall test, but a bias for threatening material on the implicit fragment completion test. While Bower's ANM provided a useful framework to stimulate research into mood and memory, and does seem to hold for depression, it does not seem to be able to deal with anxiety.

Attentional bias

The difference between anxiety and depression goes beyond memory. In anxiety, there is often a sensitivity to dangers and threats in the environment, and a fear of not being able to cope, or of panicking. It had been argued that anxiety might be associated not with memory, but with an attentional bias: a predisposition to look for and notice threats. Since this would lead to the person perceiving a more threatening world, it would be self-fulfilling. They would feel more anxious, and so search more keenly for threats. This hypothesis has been supported by evidence from a number of cognitive attentional tasks.

One of the most widely used tasks is the **emotional Stroop**, a modification of the classic **Stroop task** (in which the names of colors are printed in different colored inks, see Topic C4). In the emotional Stroop, words that are related or unrelated to a type of threat are printed in different colors, and people are asked to name the colors. The extent to which they are slowed down by the nature of the words indicates the degree of interference or attentional distraction that the words cause. Mathews and Macleod (1986) found that anxious patients took longer to name the colors of words such as CANCER, PAIN and DEATH. Watts *et al.* (1986) reported that spider phobics showed a similar increase in the time taken to name the colors of spider related words such as HAIRY, COBWEB and SPINDLY. In a **dichotic listening task**, Parkinson and Rachman (1981) asked two groups of mothers to listen to music in which they had embedded a list of words. One group of mothers had children who were in hospital, the others did not. Words that were related to sickness and hospitals were noticed more often by the mothers of the sick children.

Both of these tasks could also be influenced by response competition though: people might have no predisposition to notice the words more, but anxious people might not be able to inhibit their reactions to the threatening words. This would interfere with any other response that they have to make, such as naming the color or listening for and remembering the next word in the tape. To avoid this, Macleod and colleagues adapted an attentional task which did not measure responses to the threat cue: the **dot probe task** (*Fig. 1*; see Topic C2). In this task, people are asked to fixate on a cross displayed in the middle of a computer screen, and to press a button when they see a dot appear either above or below the cross. Before each dot appears, two words are displayed, one above and one below the cross, and people have to read out the top word. They have just enough time to do so. Dots that appear above the cross are noticed faster than dots that appear below the cross, because people are directing their attention there in order to read the top word. However, when a threatening word appears below the cross, anxious people have their attention drawn towards it, even though they do not have to read it. This slows down their reaction to dots that appear above the cross, and speeds up responses to dots that appear below the cross. Non-anxious people show the opposite pattern: they are even faster for dots above the cross, and even slower for dots below the cross. This seems to show that anxiety does indeed influence the way that we direct our attention around the world, and that being anxious can actually makes us more likely to notice things that we fear.

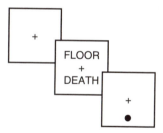

Fig. 1.   In the Attentional Dot Probe Task, people have to read out the top word from a briefly presented pair, and then press a button if they see a dot follow either of the words (or in some versions, press one of two buttons to indicate where the dot was). Half of the words that appear in both the top and bottom positions are 'threatening', and a measure of attentional bias to threatening stimuli is given by the difference in reaction times to dots that follow threat words and non-threat words. Anxious people are faster to notice dots that appear in the same location as threat words; non-anxious controls are slower.

L3 EMOTION, PERCEPTION AND PERFORMANCE

Keynotes

Optimal arousal	The Yerkes–Dodson law of arousal is that there is a certain optimal level of arousal for performance, and that lower or higher levels lead to worse performance. Emotions serve to match our arousal level to the situation, to optimize performance. Easterbrook hypothesized that arousal was linked to attention, with high levels of arousal narrowing attention so that we do not become distracted by peripheral events.
Decision making	The elaboration likelihood model (ELM) says that when we are in a good mood, we are less likely to elaborate details of a situation, but will be influenced by peripheral details (such as someone's social status). In negative moods, we are motivated to elaborate the details of the situation, and will be less biased by peripheral aspects. Oaksford and colleagues relate positive mood to a reluctance to plan our actions, leading to worse performance on tasks that require planning. Forgas's affect infusion model (AIM) predicts that mood has most effect when we are either attending to as many details as possible or when we are trying to make a rapid heuristic decision.
Mood and creativity	Problems that require conventional strategies and ideas to be put to one side in favor of unusual associations appear to benefit from being in a good mood. Isen argues that this is because positive affect leads to a more inclusive style of thinking in which ambiguities and unusual ideas are more tolerated.

Related topics	Expertise (F3)	Mood, memory and attention (L2)
	Decision making (G4)	

Optimal arousal
One of the key functions of emotion is that it acts as a motivating force, that is, it moves us to action. When we are angry or frightened, or when we are excited or interested, we are motivated to behave in particular ways. We have an increased tendency to either act aggressively towards, or to flee from, or to approach the cause of our emotion. The behavior that is motivated is appropriate to the situation, and so, the argument goes, emotions are adaptive responses to the stimuli. A very early explanation, by Yerkes and Dodson, focused on the level of physiological **arousal** associated with different emotional states. Different levels of arousal lead to different levels of performance, with very low levels of arousal and very high levels of arousal leading to poor performance on a task (due to actions being carried out too slowly or too hastily). The best or optimal performance, with actions being carried out at just the right speed, occurs at an intermediate level of arousal. This pattern is known as the 'inverted-U' curve, because as arousal increases, performance first rises, then levels off, and finally falls again (*Fig. 1*).

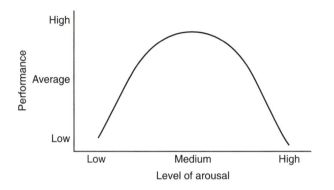

Fig. 1. An 'inverted U curve' showing the relationship between level of arousal and performance on a typical task. For easier, or more routine, tasks, the 'peak' of the U would move to the right, while for harder tasks it would move to the left.

Cognitive theory asks why and how this relationship exists. One clue from the data reported by Yerkes and Dodson is that the exact level of arousal that results in the best performance depends on the situation. Some situations require a moderately low level of arousal, some a medium level, and others a high level. Emotion could be a way of matching our arousal level to the situation, to produce the best level of performance. Difficult or challenging tasks require low arousal and concentrated effort, so might be associated with negative or depressed moods. Easy tasks that can be done very quickly benefit from high arousal, and so might be associated with positive and even anxious moods. Clinical dysfunctions occur when the relationship between situations and moods becomes maladapted, and one's arousal pattern shifts towards over-arousal (anxiety) or under-arousal (depression).

Easterbrook (1959) proposed that the relationship between arousal level and performance could be due to attention. According to his hypothesis, the role of increasing arousal is to narrow the field of **attention**, so that peripheral events are less likely to be processed or noticed. When we are in a low state of arousal, all events, peripheral or central, are equally well processed. High arousal would therefore be beneficial when we are in threatening or dangerous situations and need to concentrate on the dangerous aspect of our environment. Peripheral aspects would fall outside the narrow field of attention and would not distract us from the threat. Low arousal would be beneficial when we are in a situation where many events need to be processed, and so we need a wide field of attention to take them all in. An experiment by Heuer and Reisberg (1990) in which people are shown a series of pictures and read a story that can be either arousing or neutral tend to support this hypothesis, with the arousing story increasing subsequent memory for 'gist' items that were at the focus of attention, but decreasing memory for 'non-gist' items in the pictures.

Decision making Positive and negative moods have also been related to changes in sensitivity to information, particularly in the area of decision making and persuasion. Petty and Cacioppo's **elaboration likelihood model** (or ELM, 1986) holds that when we are in good moods, we have no motivation to change our state or to question aspects of our situation. The likelihood of the elaboration of information (the amount of effort expended on processing it) is dependent upon our mood state. When we are in bad moods, however, we are motivated to examine our situation to find out why we are unhappy or dissatisfied, so that we can take action to change the situation and make ourselves happier. Good moods are associated with **peripheral processing**, where we accept information without processing it in much detail, and bad moods with **central processing** where we focus on details of information and seek to establish whether it is relevant or not. In the context of persuasion, Mackie and Worth found that while people in a neutral mood were more influenced by a strong argument than a weak argument (because they had attended to the detail of the argument), people in a positive mood were more influenced by the status of the person making the argument (an irrelevant detail, which was peripheral to the quality of the argument). Their subsequent work (Mackie and Worth, 1989) has questioned whether these effects are entirely motivational: giving people who are in a good mood extra time increases the degree to which they attend to the content of the argument, and giving people who are in a bad mood a distracting task to do decreases the effect of the argument.

Oaksford *et al.* (1996) examined the role that mood played in reasoning by inducing volunteers into different moods and asking them to complete a version of **Wason's selection task** (see Topic G3). While 42% of those in a neutral mood chose the correct solution, only 21% of those in the negative mood, and just 9% of those in the positive mood did so. They considered the possibility that **mood-congruent retrieval** might be the reason for this result, with the positive and negative moods bringing to mind irrelevant mood-congruent information, to interfere with the selection task. Positive moods would, in non-depressed people, have more memories related to them, and so interfere more than negative moods. However, results on another reasoning task did not confirm this. The **Tower of London task** is often used to test the **planning** and control abilities of patients with head injuries, and consists of three stacks of different colored balls. The task is to move the balls from a starting position to an end position, and requires planning to avoid getting the ball that needs to be moved stuck under one that is in the right place (see *Fig. 1* in Topic F2). They found that on this task, a negative mood did not make performance worse, but a positive mood did. It seems that, as in the ELM, the happier people were, the less attention they paid to details, and the more mistakes they made.

A systematic account of the effect of emotion on decision making has been given by Forgas in his **affect infusion model** (or AIM, 1995). This assumes that the effect of any information upon a decision will be greatest when it is being processed in an appropriate manner. Forgas identifies four types of processing, each having different implications for the influence of emotion. **Direct access processing** uses well-learnt procedures for familiar problems, and **motivated processing** is directed towards coming to a particular predetermined decision. Mood state will have little effect in these two cases. **Heuristic processing** uses 'quick and dirty' strategies to avoid detailed processing, and **systematic processing** uses detailed and thorough evaluation of all aspects of a decision. Both of these will be influenced by mood, for different reasons. When people are trying to use shortcuts to come to a heuristic decision, they may use their mood state as indicative of the appropriate decision. On the other hand, when a detailed evaluation of all information is being made, memory biases may lead to mood-congruent aspects of the decision being more influential, and so lead to the infusion of affect that gives the model its name.

Mood and creativity

Isen argues that in order to come up with creative ideas, unusual associations or aspects of the situation have to be allowed to come to mind. Normally, these would have to be suppressed or inhibited to prevent us getting distracted, and to ensure that we recall information that is relevant to the task in hand; but when conventional strategies fail to provide a solution, or when a novel solution is required, these unusual aspects may provide the key. In one study (Isen *et al.*, 1987), she asked medical students to solve Dunker's candle problem (see Topic F1), in which you are given a candle and a box of tacks, and asked to fix the candle to the wall so it can be lit without dripping wax on the floor. The problem can only be solved if an unusual use is thought of for the box that the tacks are held in: it can be tacked to the wall, thus being used as a tray to both hold the candle and catch the wax. Before giving the students the problem, she gave half of them a small bar of candy as a reward for helping. These students, whose mood was made more positive than a control group who received no reward, were more likely to solve the problem, and to do so faster. It would seem that the positive mood increased their ability to consider unusual associations. This result is also in line with the increased sensitivity to peripheral information predicted by the ELM and by Easterbrook.

REFERENCES

Allport, D.A., Antonis, B. and Reynolds, P. (1972) On the division of attention: A disproof of the single channel hypothesis. *Quarterly Journal of Experimental Psychology, 24,* 225–235.

Anderson, J.R. (1978) Arguments concerning representations from mental imagery. *Psychological Review, 89,* 249–277.

Anderson, J.R. (1983) *The Architecture of Cognition.* Harvard University Press, Cambridge, MA.

Anderson, J.R. (2000) *Cognitive Psychology and its Implications,* 5th Edn. Worth, New York.

Andrade, J. and Meudell, P.R. (1993) Is spatial information encoded automatically in memory? *Quarterly Journal of Experimental Psychology, 46A,* 365–375.

Andrade, J., Kemps, E., Wernier, Y., May, J. and Szmalec, A. (2002) Insensitivity of visual short-term memory to irrelevant visual information. *Quarterly Journal of Experimental Psychology, 55(3),* 753–774.

Antos, S.J. (1979) Processing facilitation in a lexical decision task. *Journal of Experimental Psychology: Human Perception and Performance, 5,* 527–545.

Atkinson, R.C. and Shiffrin, R.M. (1968) Human memory: a proposed system and control processes. In K.W. Spence and J.D. Spence (Eds), *The Psychology of Learning and Motivation,* Vol. 2. Academic Press, New York.

Baars, B.J. (1988) *A Cognitive Theory of Consciousness.* Cambridge University Press, New York.

Baddeley, A.D. (1966a) Short-term memory for word sequences as a function of acoustic, semantic and formal similarity. *Quarterly Journal of Experimental Psychology, 18,* 362–365.

Baddeley, A.D. (1966b) The influence of acoustic and semantic similarity on long-term memory for word sequences. *Quarterly Journal of Experimental Psychology, 18,* 302–309.

Baddeley, A.D. (1986) *Working Memory.* Oxford University Press, Oxford.

Baddeley, A.D. (1993) Working memory and conscious awareness. In A. F. Collins, S. E. Gathercole, M. A. Conway and P. E. Morris (Eds), *Theories of Memory.* Lawrence Erlbaum Associates Ltd., Hove, UK, pp. 11–28.

Baddeley, A.D. (1996) Exploring the central executive. *Quarterly Journal of Experimental Psychology, 49,* 5–28.

Baddeley, A. (2000) The episodic buffer: a new component of working memory? *Trends in Cognitive Science, 4(11),* 417–423.

Baddeley, A.D. and Andrade, J. (2000) Working memory and the vividness of imagery. *Journal of Experimental Psychology: General, 129(1),* 126–145.

Baddeley, A.D. and Hitch, G.J. (1974) Working memory. In G. Bower (Ed.), *The Psychology of Learning and Motivation.* Academic Press, New York, pp. 47–89.

Baddeley, A.D. and Lieberman, K. (1980) Spatial working memory. In R.S. Nickerson (Ed.), *Attention and Performance VIII.* LEA, Hillsdale, NJ, pp. 521–539.

Baddeley, A.D. and Wilson, B.A. (1994) When implicit learning fails: Amnesia and the problem of error elimination. *Neuropsychologia, 32,* 53–68.

Baddeley, A.D., Thomson, N. and Buchanan, M. (1975) Word length and the structure of short-term memory. *Journal of Verbal Learning and Verbal Behavior, 14,* 575–589.

Baddeley, A.D., Logie, R., Bressi, S., Della Sala, S. and Spinnler, H. (1986) Dementia and working memory. *Quarterly Journal of Experimental Psychology, 38A,* 603–618.

Baddeley, A.D., Emslie, H., Kolodny, J. and Duncan, J. (1998a) Random generation and the executive control of working memory. *Quarterly Journal of Experimental Psychology, 51A(4),* 819–852.

Baddeley, A., Gathercole, S. and Papagno, C. (1998b) The Phonological loop as a language learning device. *Psychological Review, 105,* 158–173.

Bahrick, H.P. (1984) Semantic memory content in permastore: Fifty years of memory for Spanish learned in school. *Journal of Experimental Psychology: General, 113,* 1–29.

Barnard, P.J. (1985) Interacting cognitive subsystems: A psycholinguistic approach to short-term memory. In A. Ellis (Ed.), *Progress in the Psychology of Language,* Vol. 2. Lawrence Erlbaum Associates Ltd, London, pp. 197–258.

Barsalou, L.W. (1982) Context-independent and context-dependent information in concepts. *Memory and Cognition, 10,* 82–93.

Bartlett, F.C. (1932) *Remembering.* Cambridge University Press, Cambridge.

Bauer, R.M. (1984) Automatic recognition of names and faces in prosopagnosia: A neuropsychological application of the guilty knowledge test. *Neuropsychologia, 22,* 457–469.

Beck, A.T. (1967) *Depression: Clinical, Experimental and Theoretical Aspects.* Harper and Row, New York.

Berko, J. (1958) The child's learning of English morphology. *Word, 14,* 150–177.

Berlin, B. and Kay, P. (1969) *Basic Color Terms: Their Universality and Evolution.* University of California Press, Berkeley and Los Angeles.

Biederman, I. (1987) Recognition-by-components: A theory of human image understanding. *Psychological Review, 94,* 115–147.

Bierwisch, M. (1967) Some semantic universals of German adjectivals. *Foundations of Language, 3,* 1–36.

Billingsley, R.L., Smith, M.L. and McAndrews, M.P. (2002) Developmental patterns in priming and familiarity in explicit recollection. *Journal of Experimental Child Psychology, 82,* 251–277.

Bishop, D.V.M., North, T. and Donlan, C. (1996) Nonword repetition as a behavioural marker for inherited language impairment: Evidence from a twin study. *Journal of Child Psychology and Psychiatry, 37,* 391–403.

Block, N. (1995) On a confusion of a function of consciousness. *Behavioral and Brain Sciences, 18,* 227–287.

Bower, G.H. (1981) Mood and Memory. *American Psychologist*, 36, 129–148.

Bower, G.H., Gilligan, S.G. and Monteiro, K.P. (1981) Selectivity of learning caused by affective states. *Journal of Experimental Psychology: General*, 110, 451–473.

Bransford, J.D. and Johnson M.K. (1972) Contextual prerequisites for understanding: some investigations of comprehension and recall. *Journal of Verbal Learning and Verbal Behavior*, 11, 717–726.

Broadbent, D.E. (1958) *Performance and Communication*. Pergamon Press, Oxford.

Brooks, L.R. (1967) The suppression of visualization by reading. *Quarterly Journal of Experimental Psychology*, 19, 289–299.

Brown, I.D., Tickner, A.H. and Simmonds, D.C.V. (1969) Interference between concurrent tasks of driving and telephoning. *Journal of Applied Psychology*, 53, 419–424.

Brown, R. (1973) *A First Language: The Early Stages*. Harvard University Press, Cambridge, MA.

Brown, R. and Kulik, J. (1977) Flashbulb memories. *Cognition*, 5, 73–99.

Brown, R. and McNeill, D. (1966) The 'tip of the tongue' phenomenon. *Journal of Verbal Learning and Verbal Behavior*, 5, 325–327.

Bruce, V. and Young, A.W. (1986) Understanding face recognition. *British Journal of Psychology*, 77, 305–327.

Burgess, N. and Hitch, G.J. (1999) Memory for serial order: A network model of the phonological loop and its timing. *Psychological Review*, 106, 551–581.

Burton, A.M., Bruce, V. and Hancock, P.J.B. (1999) From pixels to people: A model of familiar face recognition. *Cognitive Science*, 23, 1–31.

Carraher, T.N., Carraher, D.W. and Schliemann, A.D. (1985) Mathematics in the streets and schools. *British Journal of Developmental Psychology*, 3, 21–29.

Carroll, J.B. (1993) *Human Cognitive Abilities*. Cambridge University Press, Cambridge.

Carroll, J.B. and Casagrande, J.B. (1958) The function of language classification in behavior. In E.E. Maccoby, T. Newcomb and E.L. Hartley (Eds), *Readings in Social Psychology*, 3rd Edn. Holt, Reinhart & Winston, New York, pp 18–31.

Case, R., Demetriou, A., Platsidou, M. and Kazi, S. (2001) Integrating concepts and tests of intelligence from the differential and developmental traditions. *Intelligence*, 29, 307–336.

Cattell, R.B. (1971) *Abilities, Their Structure, Growth and Action*. Houghton Miflin, New York.

Ceci, S.J. and Bronfenbrenner, U. (1985) Don't forget to take the cup cakes out of the oven: Prospective memory, strategic time monitoring and context. *Child Develop.*, 56, 152–164.

Chalmers, D. (1996) *The Conscious Mind*. Oxford University Press, Oxford.

Chambers, D. and Reisberg, D. (1985) Can mental images be ambiguous? *Journal of Experimental Psychology: Human Perception and Performance*, 11, 317–328.

Chapman, L.J. and Chapman, J.P. (1959) Atmosphere effect re-examined. *Journal of Experimental Psychology*, 58, 220–226.

Chase, W.G. and Simon, H.E. (1973) Perception in Chess. *Cognitive Psychology*, 4, 55–81.

Cheng, P.W. and Holyoak, K.J. (1985) Pragmatic reasoning schemas. *Cognitive Psychology*, 17, 391–416.

Cherry, E.C. (1953) Some experiments on the recognition of speech with one and two ears. *Journal of the Acoustical Society of America*, 25, 975–979.

Chomsky, N. (1957) *Syntactic Structures*. Mouton, The Hague.

Chomsky, N. (1965) *Aspects of the Theory of Syntax*. MIT Press, Cambridge, MA.

Christensen, I.P., Wagner, H.L. and Halliday, M.S. (2001). *Psychology*. BIOS Scientific Publishers Ltd., Oxford.

Clark, H.H. (1996) *Using Language*. University of Chicago Press, Chicago.

Clark, H.H. and Sengul, C.J. (1979) In search of referents for nouns and pronouns. *Memory and Cognition*, 7, 35–41.

Clifton, C. and Ferreira, F. (1987) Discourse structure and anaphora: some experimental results. In M. Coltheart (Ed.), *Attention and Performance 12*. Erlbaum, London.

Colle, H.A. and Welsh, A. (1976) Acoustic masking in primary memory. *Journal of Verbal Learning and Verbal Behavior*, 15, 17–32.

Collins, A.M. and Loftus, E.F. (1975) A spreading-activation theory of semantic processing. *Psychological Review*, 82, 407–428.

Collins, A.M. and Quillian, M.R. (1969) Retrieval time from semantic memory. *Journal of Verbal Learning and Verbal Behavior*, 9, 240–248.

Conrad, R. and Hull, A.J. (1964) Information, acoustic confusion and memory span. *British Journal of Psychology*, 55, 429–432.

Conway, A.R.A., Cowan, N. and Bunting, M.F. (2001) The cocktail party phenomenon revisited: The importance of working memory capacity. *Psychonomic Bulletin and Review*, 8, 331–335.

Conway, M.A. *et al.* (1994) The formation of flashbulb memories. *Memory and Cognition*, 22, 326–343.

Cooper, R. and Shallice, T. (2000) Contention scheduling and the control of routine activities. *Cognitive Neuropsychology*, 17, 297–338.

Cosmides, L. (1989) The logic of social exchange: has natural selection shaped how humans reason. Studies with the Wason selection task. *Cognition*, 31, 187–276.

Cowan, N. (1993) Activation, attention, and short-term memory. *Memory and Cognition*, 21, 162–167.

Craik, F.I.M. and Lockhart, R.S. (1972) Levels of processing: A framework for memory research. *Journal of Verbal Learning and Verbal Behavior*, 11, 671–684.

Craik, F.I.M. and Watkins, M.J. (1973) The role of rehearsal in short-term memory. *Journal of Verbal Learning and Verbal Behavior*, 12, 599–607.

Crick, F. and Koch, C. (1990) Towards a neurobiological theory of consciousness. *Seminars in the Neurosciences* 2, 263–275.

Csikszentmihalyi, M. (1988) Society, culture and person: a systems view of creativity. In R.J. Sternberg (Ed.), *The Nature of Creativity: Contemporary Psychological Perspectives*. Cambridge University Press, Cambridge, pp. 325–339.

Curiel, J.M. and Radvansky, G.A. (2002) Mental maps in memory retrieval and comprehension. *Memory*, 10, 113–126.

Curtis, M.E. (1977) *Genie: A Linguistic Study of a Modern Day 'Wild Child'*. Academic Press, New York.

Daneman, M. and Carpenter, P.A. (1980) Individual differences in working memory and reading. *Journal of Verbal Learning and Verbal Behavior*, 19, 450–466.

Davis, K. (1947) Final note on a case of extreme social isolation. *American Journal of Sociology*, 52, 432–437.

Deeprose, C., Andrade, J., Varma, S. and Edwards, N. (2004) Unconscious learning during surgery with propofol anaesthesia. *Br. J. Anaesthesia* (in press).

Dennett, D. (1991) *Consciousness Explained*. Little, Brown & Co., Boston, MA.

Denny, E.B. and Hunt, R.R (1992) Affective valence and memory in depression: dissociation of recall and fragment completion. *Journal of Abnormal Psychology*, 101, 575–580.

Deutsch, F.A. and Deutsch, D. (1963) Attention: Some theoretical considerations. *Psychological Review*, 70, 80–90.

Dickstein, L.S. (1978) The effect of figure on syllogistic reasoning. *Memory and Cognition*, 6, 76–83.

Dickstein, L.S. (1981) The meaning of conversion in syllogistic reasoning. *Bulletin of the Psychonomic Society*, 18, 135–138.

DiLollo, V., Hansen, D. and McIntyre, J.S. (1983) Initial stages of visual information processing in dyslexia. *Journal of Experimental Psychology: Human Perception and Performance*, 9, 923–935.

Driver, J. (1996) Enhancement of selective listening by illusory mislocation of speech sounds due to lip-reading. *Nature*, 381, 66–68.

Duncan, J. (1984) Selective attention and the organization of visual information. *Journal of Experimental Psychology: General*, 113, 501–517.

Duncan, J. and Humphreys, G.W. (1992) Beyond the search surface: Visual search and attentional engagement. *Journal of Experimental Psychology: Human Perception and Performance*, 18, 578–588.

Duncker, K. (1945) On Problem Solving. *Psychological Monographs*, 58, No. 270.

Easterbrook, J.A. (1959) The effect of emotion on cue utilization and the organization of behavior. *Psychological Review*, 66, 183–201.

Ebbinghaus, H. (1885) *Über das Gedächtnis*. Dunker, Leipzig.

Ekstrand, B.R., Barrett, T.R., West, J.M. and Maier, W.G. (1977) The effect of sleep on human memory. In Drucker-Colin, R. and McGaugh, J. L. (Eds), *Neurobiology of Sleep and Memory*. Academic Press, New York.

Engle, R.W. (2002) Working memory capacity as executive attention. *Current Directions in Psychological Science*, 11, 19–23.

Engle, R.W., Tuholski, S.W., Laughlin, J.E. and Conway, A.R.A. (1999) Working-memory, short-term memory, and general fluid intelligence: A latent variable approach. *Journal of Experimental Psychology: General*, 128, 309–331.

Ericcson, K.A., Chase, W.G. and Faloon, S. (1980) Acquisition of a memory skill. *Science*, 208, 201–204.

Escalona, S.K. (1973) Basic modes of social interaction: their emergence and patterning during the first two years of life. *Merrill-Palmer Quarterly*, 19, 205–232.

Evans, J.St.B.T. (1984) Heuristic and analytic processes in reasoning. *British Journal of Psychology*, 75, 457–468.

Evans, J.St.B.T., Barston, J. and Pollard, P. (1983) On the conflict between logic and belief in syllogistic reasoning. *Memory and Cognition*, 11, 295–306.

Farah, M.J. (1985) Psychophysical evidence for a shared representational medium for visual images and percepts. *Journal of Experimental Psychology: General*, 114, 91–103.

Farah, M.J., Hammond, K.M., Levine, D.N. and Calvanio, R. (1988) Visual and spatial mental imagery: Dissociable systems of representation. *Cognitive Psychology*, 20, 439–462.

Fawcett, A.J. and Nicolson, R.I. (1995) Persistence of phonological awareness deficits in older children with dyslexia. *Reading and Writing: An Interdisciplinary Journal*. 7, 361–376.

Fawcett, A.J., Nicolson, R.I. and Dean, P. (1996) Impaired performance of children with dyslexia on a range of cerebellar tasks. *Annals of Dyslexia*, 46, 259–283.

Finke, R.A. (1995) Creative Insight and preinventive forms. In R.J. Sternberg and J.E. Davidson (Eds), *The Nature of Insight*. Cambridge University Press, Cambridge, pp. 255–280.

Finke, R.A. and Pinker, S. (1982) Spontaneous imagery scanning in mental extrapolation. *Journal of Experimental Psychology: Learning, Memory and Cognition*, 8, 142–147.

Finn, P.R., Justus, A., Mazas, C. and Steinmetz, J.E. (1999) Working memory, executive processes and the effects of alcohol on Go/No-Go learning: testing a model of behavioral regulation and impulsivity. *Psychopharmacology*, 146, 465–472.

Flynn, J.R. (1987) Massive IQ gains in 14 nations: what IQ tests really measure. *Psychological Bulletin*, 101, 171–191.

Fodor, J.A. (1976) *The Language of Thought*. Harvester, Hassocks, Sussex.

Forgas, J.P. (1995) Mood and judgement: The affect infusion model (AIM). *Psychological Bulletin*, 17, 39–66.

Forster, K.I. (1979) Levels of processing and the structure of the language processor. In W.E. Cooper and E.C.T. Walker (Eds), *Sentence Processing Psycholinguistic Studies presented to Merrill Garrett*. Lawrence Erlbaum, Hillsdale, NJ.

Frith, U. (1997) Brain, mind and behaviour in dyslexia. In C. Hulme and M. Snowling (Eds), *Dyslexia: Biology, Cognition and Intervention*. Whurr, London.

Gardiner, J.M. and Java, R.I. (1993) Recognising and remembering. In A.F. Collins, S.E. Gathercole, M.A. Conway and P.E. Morris (Eds), *Theories of Memory*. Lawrence Erlbaum Associates Ltd., Hove, UK.

Gardner, H. (1983) *Frames of Mind: The Theory of Multiple Intelligences*. Basic Books, New York.

Gardner, H. (1993) *Creating Minds*. Basic Books, New York.

Gardner, R.A. and Gardner, B.T. (1969) Teaching sign language to a chimpanzee. *Science*, 165, 664–672.

Garnham, A. (1985) *Psycholinguistics*. Methuen, London.

Garrett, M.F. (1980) Levels of processing in sentence production. In B. Butterworth (Ed.), *Language Production*, Vol. 1: *Speech and Talk*. Academic Press, London, pp. 177–210.

Gathercole, S.E. and Baddeley, A.D. (1989) Evaluation of the role of phonological STM in the development of vocabulary in children: A longitudinal study. *Journal of Memory and Language*, 28, 200–215.

Gentner, D. and Gentner, D.R. (1983) Flowing waters and teeming crowds: mental models of electricity. In D.R. Gentner and A.L. Stevens (Eds), *Mental Models*. Lawrence Erlbaum, Hillsdale, NJ, pp. 99–129.

Gernsbacher, M.A. (1985) Surface information loss in comprehension. *Cognitive Psychology*, 17, 324–363.

Gick, M.L. and Holyoak, K.J. (1980) Analogical problem solving. *Cognitive Psychology*, 12, 306–355.

Glenberg, A.M., Meyer, M. and Lindem, K. (1987) Mental models contribute to foregrounding during test comprehension. *Journal of Memory and Language*, 26, 69–83.

Godden, D.R. and Baddeley, A.D. (1975) Context-dependent memory in two natural environments: On land and under water. *British Journal of Psychology*, 66, 325–331.

Goldin-Meadow, S. and Mylander, C. (1998). Spontaneous sign systems created by deaf children in two cultures. *Nature, 391*, 279–281

Golinkoff, R.M., Hirsh-Pasek, K., Cauley, K. and Gordon, L. (1987) The eyes have it: lexical and semantic comprehension in a new paradigm. *Journal of Child Language, 14*, 23–46.

Greenfield, P.M. and Smith, J.H. (1976) *The Structure of Communication in Early Language Development*. Academic Press, New York.

Gregory, R.L. (1980) Perceptions as hypotheses. *Philosophical Transactions of the Royal Society of London, Series B, 290*, 181–197.

Grice, H.P. (1975) Logic and conversation. In P. Cole and J.L. Morgan (Eds), *Syntax and Semantics 3: Speech Acts*. Seminar Press, New York.

Guilford, J.P. (1967) *The Nature of Human Intelligence*. McGraw-Hill, New York.

Gunter, B., Berry, C. and Clifford, B.R. (1981) Proactive interference effects with television news items: Further evidence. *Journal of Experimental Psychology: Human Learning and Memory, 7*, 480–487.

Hasher, L. and Zacks, R.T. (1979) Automatic and effortful processes in memory. *Journal of Experimental Psychology: General, 108*, 356–388.

Haviland, S.E. and Clark, H.H. (1974) What's new? Acquiring new information as a process in comprehension. *Journal of Verbal Learning and Verbal Behavior, 13*, 512–521.

Heuer, F. and Reisberg, D. (1990) Vivid memories of emotional events – the accuracy of remembered minutiae. *Memory and Cognition, 18*, 496–506.

Hockett, C.F. (1966) The problem with universals in language. In J.H. Greenberg (Ed.), *Universals of Language*, 2nd Edn. MIT Press, Cambridge, MA, pp 1–29.

Hoffman, C., Lau, L. and Johnson, D.R. (1986) The linguistic relativity of person cognition: An English–Chinese comparison. *Journal of Personality and Social Psychology, 51*, 1097–1105.

Homa, D., Sterling, S. and Trepel, L. (1981) Limitations of exemplar-based generalization and the abstraction of categorical information. *Journal of Experimental Psychology: Human Learning and Memory, 7*, 418–439.

Howell, P. and Powell, D.J. (1987) Delayed auditory feedback with delayed sounds varying in duration. *Perception and Psychophysics, 42*, 166–172.

Humphreys, G.W. and Bruce, V. (1989) *Visual Cognition*. Lawrence Erlbaum Associates, Hove.

Hunt, E. (1985) The correlates of intelligence. In D.K. Detterman (Ed.), *Current Topics in Human Intelligence*, Vol. 1. Ablex, Norwood, NJ.

Hunt, E. and Agnoli, F. (1991) The Whorfian hypothesis: a cognitive psychology perspective. *Psychological Review, 98*, 377–389.

Huppert, F.A. and Beardsall, L. (1993) Prospective memory impairment as an early indicator of dementia. *Journal of Clinical and Experimental Neuropsychology, 15*, 805–821.

Isen, A., Daubman, K.A. and Nowicki, G.P. (1987) Positive affect facilitates creative problem solving. *Journal of Personality and Social Psychology, 52*, 1122–1131.

Ittelson, W.H. (1951) Size as a cue to distance: Static localisation. *American Journal of Psychology, 64*, 54–67.

James, W. (1890/1918) *The Principles of Psychology*, Vol. I. Macmillan and Co. Ltd, London.

Jensen, A.R. (1982) Reaction time and psychometric g. In H.J. Eysenck (Ed.), *A Model for Intelligence*. Springer–Verlag, Berlin, pp 93–132.

Johnson, J. and Newport, E. (1989) Critical period efforts in second language learning: The influence of maturational state on the acquisition of English as a second language. *Cognitive Psychology, 21*, 60–99.

Johnson-Laird, P.N. (1983) *Mental Models*. Cambridge University Press, Cambridge.

Johnson-Laird, P.N. and Steedman, M. (1978) The psychology of syllogisms. *Cognitive Psychology, 10*, 64–99.

Johnson-Laird, P.N. and Wason, P.C. (1970) Insight into a logical relation. *Quarterly Journal of Experimental Psychology, 22*, 49–61.

Johnston, W.A. and Heinz, S.P. (1978) Flexibility and capacity demands of attention. *Journal of Experimental Psychology: General, 107*, 420–435.

Jones, D.M., Alford, D., Bridges, A., Tremblay, S. and Macken, B. (1999) Organizational factors in selective attention: The interplay of acoustic distinctiveness and auditory streaming in the irrelevant sound effect. *Journal of Experimental Psychology: Learning, Memory, and Cognition, 25*, 464–473.

Juola, J.F., Bowhuis, D.G., Cooper, E.E. and Warner, C.B. (1991) Control of attention around the fovea. *Journal of Experimental Psychology: Human Perception and Performance, 15*, 315–330.

Just, M.A., Carpenter, P.A. and Masson, M.E.J. (1982) *What eye fixations tell us about speed reading and skimming*. Eye-Lab Technical report: Carnegie Mellon University.

Kahneman, D. (1973) *Attention and Effort*. Prentice Hall, Englewood Cliffs, NJ.

Kahneman, D. and Tversky, A. (1973) On the psychology of prediction. *Psychological Review, 80*, 237–251.

Kanizsa, G. (1976) Subjective contours. *Scientific American, 234(4)*, 48–52.

Kaplan, C.A. and Simon, H.A. (1990) In search of insight. *Cognitive Psychology, 22*, 374–419.

Kemps, E. (2001) Complexity effects in visuo–spatial working memory: Implications for the role of long-term memory. *Memory, 9*, 13–27.

Kintsch, W. (1988) The role of knowledge in discourse comprehension: A Construction–Integration model. *Psychological Review, 95*, 163–182.

Klayman, J. and Ha, Y-W. (1987) Confirmation, disconfirmation and information in hypothesis testing. *Psychological Review, 94*, 211–228.

Kleiman, G.M. (1975) Speech recoding in reading. *Journal of Verbal Learning and Verbal Behavior, 14*, 323–339.

Kosslyn, S.M. (1994) *Image and Brain: The Resolution of the Imagery Debate.* MIT Press, Cambridge, MA.

Kosslyn, S.M., Ball, T.M. and Reiser, B.J. (1978) Visual images preserve metric spatial information: Evidence from studies of image scanning. *Journal of Experimental Psychology: Human Perception and Performance, 4*, 47–60.

Kunst-Wilson, W.R. and Zajonc, R.B. (1980) Affective discrimination of stimuli that cannot be recognized. *Science, 207*, 557–8.

Kvavilashvili, L. (1987) Remembering intention as a distinct form of memory. *British Journal of Psychology, 78*, 507–518.

Kyllonen, P.C. and Christal, R.E. (1990) Reasoning ability is (little more than) working memory capacity?! *Intelligence, 14*, 389–433.

Landau, B. and Gleitman, L.R. (1985) *Language and Experience: Evidence from the Blind Child.* Harvard University Press, Cambridge, MA.

Larkin, J.H., McDermott, J., Simon, D.P. and Simon, H.A. (1980) Expert and novice performance in solving physics problems. *Science, 208*, 1335–1342.

Lazarus, R.S. (1982) Thoughts on the relations between emotion and cognition. *American Psychologist, 37*, 1019–1024.

Lee, B. (1950) Effects of delayed speech feedback. *Journal of the Acoustical Society of America, 22*, 824–826

Lesgold, A. (1988) Problem Solving. In R.J. Sternberg and E.E. Smith (eds) *The Psychology of Human Thought.* Cambridge University Press, New York, pp. 188–316.

Lewandowsky, S. and Murdock, B.B. (1989) Memory for serial order. *Psychological Review, 96*, 25–57.

Liberman, A. and Mattingly, I. (1985) The motor theory of speech perception revisited. *Cognition, 21*, 1–36.

Lisker, L. and Abramson, A. (1964) A cross-language study of voicing in initial stops: acoustical measurement. *Word, 20*, 384–422.

Lloyd, G.G. and Lishman, W.A. (1975) Effect of depression on the speed of recall of pleasant and unpleasant experiences. *Psychological Medicine, 5*, 173–180.

Loehlin, J.D., Vandenberg, S.G. and Osborne, R.T. (1973) Blood group genes and Negro–white ability differences. *Behavior Genetics, 3*, 263–277.

Loftus, E.F. (1993) The reality of repressed memories. *American Psychologist, 48*, 518–537.

Loftus, E.F. and Palmer, J.C. (1974) Reconstruction of automobile destruction: An example of the interaction between language and memory. *Journal of Verbal Learning and Verbal Behavior, 13*, 585–589.

Logan, G.D. (1988) Toward an instance theory of automatization. *Psychological Review, 95*, 492–527.

Logie, R.H. (1995) *Visuo–spatial Working Memory.* Psychology Press, Hove, UK.

Logie, R.H., Zucco, G.M. and Baddeley, A.D. (1990) Interference with visual short-term memory. *Acta Psychologica, 75*, 55–74.

Logie, R.H., Della Sala, S., Wynn, V. and Baddeley, A.D. (2000) Visual similarity effects in immediate verbal serial recall. *Quarterly Journal of Experimental Psychology, 53*, 626–646.

Lovatt, P.J., Avons, S.E. and Masterson, J. (2000) The word-length effect and disyllabic words. *Quarterly Journal of Experimental Psychology, 53A*, 1–22.

Lovegrove, W. (1994) Visual deficits in dyslexia: Evidence and implications. In A.J. Fawcett and R.I. Nicolson (Eds) *Dyslexia in Children: Multidisciplinary Perspectives.* Harvester Wheatsheaf, Hemel Hempstead.

Luchins, A.S. (1942) Mechanization in problem solving. *Psychological Monographs, 54, No. 248.*

Mackie, D.M. and Worth, L.T. (1989) Processing deficits and the mediation of positive affect in persuasion. *Journal of Personality and Social Psychology, 57*, 27–40.

Maier, N.R.F. (1931) Reasoning in humans. II: The solution of a problem and its appearance in consciousness. *Journal of Comparative Psychology, 12*, 181–194.

Malt, B.C. (1989) An on-line investigation of prototype and exemplar strategies in classification. *Journal of Experimental Psychology: Learning, Memory, and Cognition, 15*, 539–555.

Mandler, G., Nakamura, Y. and Van Zandt, B.J.S. (1987) Nonspecific effects of exposure on stimuli that cannot be recognized. *Journal of Experimental Psychology: Learning, Memory, and Cognition, 13*, 646–648.

Manktelow, K.I. and Over, D.E. (1991) Social roles and utilities in reasoning with deontic conditionals. *Cognition, 39*, pp. 85–105.

Maquet, P. *et al.* (2000) Experience-dependent changes in cerebral activation during human REM sleep. *Nature Neuroscience, 3*, 831–836.

Marcel, A.J. (1983) Conscious and unconscious perception: Experiments on visual masking and word recognition. *Cognitive Psychology, 15*, 197–237.

Marcel, A.J. (1988) Phenomenal experience and functionalism. In A.J. Marcel and E. Bisiach. *Consciousness in Contemporary Science.* Oxford University Press, Oxford, pp. 121–158.

Marr, D. (1982) *Vision: A Computational Investigation into the Human Representation and Processing of Visual Information.* W. H. Freeman, San Francisco.

Marr, D. and Nishihara, K. (1978) Representation and recognition of the spatial organisation of three-dimensional shapes. *Philosophical Transactions of the Royal Society of London, Series B, 200*, 269–294.

Marshall, J.C. and Halligan, P.W. (1988) Blindsight and insight in visuo–spatial neglect. *Nature, 336*, 766–767.

Marshall, J.C. and Newcombe, F. (1973) Patterns of paralexia: a psycholinguistic approach. *Journal of Psycholinguistic Research, 2*, 175–200.

Marslen-Wilson, W.D. (1987) Functional parallelism in spoken word recognition. *Cognition, 25*, 71–102.

Marslen-Wilson, W.D. and Tyler, L.K. (1980) The temporal structure of spoken language and understanding. *Cognition, 8*, 1–71.

Massaro, D.W. (1998) *Processing Talking Faces: From Speech Perception to a Behavioural Principle.* MIT Press, Cambridge, MA.

Mathews, A., May, J., Mogg, K. and Eysenck, M.W. (1990) Attentional bias in anxiety: Selective search or defective filtering? *Journal of Abnormal Psychology*, *99*, 166–173.

Mathews, A.M. and Macleod, C. (1986) Discrimination of threat cues without awareness in anxiety states. *Journal of Abnormal Psychology*, *95*, 131–138.

Matt, G.E., Vacquez, C. and Campbell, W.K. (1992) Mood congruent recall of affectively toned stimuli: a meta-analytic review. *Clinical Psychology Review*, *12*, 227–255.

McConkie, G.W. (1979) On the role and control of eye movements in reading. In P.A. Kolers, M.E. Wrolstad, and H. Bouma (Eds), *Processing of Visible Language*, Vol. 1. Plenum, New York.

McGurk, H. and MacDonald, J. (1976) Hearing eyes and seeing voices. *Nature*, *264*, 746–748.

Melzack, R. (1992) Phantom limbs. *Scientific American*, 90–96, April.

Metcalfe, J. and Fisher, R.P. (1986) The relation between recognition memory and classification learning. *Memory and Cognition*, *14*, 164–173.

Meyer, D.E. and Schvaneveldt, R.W. (1971) Facilitation in recognizing pairs of words; evidence of a dependence between retrieval operations. *Journal of Experimental Psychology*, *90*, 227–235.

Meyer, D.E., Schvaneveldt, R.W. and Ruddy, M.G. (1974) Functions of graphemic and phonemic codes in visual word recognition. *Memory and Cognition*, *2*, 309–321.

Miller, G.A. (1956) The magic number seven plus or minus two: Some limits on our capacity to process information. *Psychological Review*, *63*, 81–97.

Milne, R.W. (1982) Predicting garden path sentences. *Cognitive Science*, *6*, 349–373.

Milner, B. (1966) Amnesia following operation on the temporal lobes. In C.W.M. Whitty and O.L. Zangwill (Eds), *Amnesia*. Butterworths, London, pp. 109–133.

Minami, H. and Dallenbach, K.M. (1946) The effect of activity upon learning and retention in the cockroach. *American Journal of Psychology*, *59*, 1–58.

Miyake, A., Friedman, N.P., Emerson, M.J., Witzki, A.H., Howerter, A. and Wager, T.D. (2000) The unity and diversity of executive functions and their contribution to complex 'frontal lobe' tasks: A latent variable analysis. *Cognitive Psychology*, *41*, 49–100.

Moray, N. (1959) Attention in dichotic listening: Affective cues and the influence of instructions. *Quarterly Journal of Experimental Psychology*, *11*, 56–60.

Morris, C.D., Bransford, J.D. and Franks, J.J. (1977) Levels of processing versus transfer appropriate processing. *Journal of Verbal Learning and Verbal Behavior*, *16*, 519–533.

Morrison, R.E. (1984) Manipulation of stimulus onset delay in reading: evidence for parallel programming of saccades. *Journal of Experimental Psychology: Human Perception and Performance*, *10*, 667–682.

Morton, J. (1969) Interaction of information in word recognition. *Psychological Review*, *76*, 165–178.

Morton, N. and Morris, R.G. (1995) Image transformation dissociated from visuospatial working memory. *Cognitive Neuropsychology*, *12*, 767–791.

Murphy, S.T. and Zajonc, R.B. (1993) Affect, cognition, and awareness: Affective priming with optimal and suboptimal stimulus exposures. *Journal of Personality and Social Psychology*, *64*, 723–739.

Murray, D.J. (1967) The role of speech responses in short-term memory. *Canadian Journal of Psychology*, *21*, 263–276.

Murray, H.G. and Denney, J.P. (1969) Interaction of ability level and interpolated activity in human problem solving. *Psychological Reports*, *24*, 271–276.

Nagel, T. (1974) What is it like to be a bat? *Philosophical Review*, *83*, 435–450.

Navon, D. (1977) Forest before trees: The precedence of global features in visual perception. *Cognitive Psychology*, *9*, 353–383.

Navon, D. and Gopher, D. (1979) On the economy of the human information processing system. *Psychological Review*, *86*, 214–255.

Neisser, U. (1964) Visual search. *Scientific American*, *210*, 94–102.

Neisser, U. (1967) *Cognitive Psychology*. Appleton–Century–Crofts, New York.

Neisser, U. and Becklen, R. (1975) Selective looking: Attending to visually specified events. *Cognitive Psychology*, *7*, 480–494.

Nettlebeck, T. (1987) Inspection time and intelligence. In P.A. Vernon (Ed.), *Speed of Information Processing and Intelligence*. Ablex, Norwood, NJ.

Newell, A. (1990) *Unified Theories of Cognition*. Harvard University Press, Cambridge, MA.

Newell, A. and Simon, H.A. (1972) *Human Problem Solving*. Prentice–Hall, New Jersey.

Newport, E. (1990) Maturational constraints on language learning. *Cognitive Science*, *14*, 11–28.

Nicolson, R.I. and Fawcett, A.J. (1990). Automaticity: A new framework for dyslexia research? *Cognition*, *35*, 159–182.

Nisbett R.E. and Wilson T. D. (1977) The halo effect: evidence for unconscious alteration of judgments. *Journal of Personality and Social Psychology*, *35*, 250–256.

Nissen, M.J. and Bullemer, P. (1987) Attentional requirements of learning: Evidence from performance measures. *Cognitive Psychology*, *19*, 1–32.

Norman, D.A. (1968) Toward a theory of memory and attention. *Psychological Review*, *75(6)*, 522–536.

Norman, D.A. and Shallice, T. (1986) Attention to action: Willed and automatic control of behavior. In R.J. Davidson, G.E. Schwartz and D. Shapiro (Eds), *Consciousness and Self-Regulation*. Plenum, New York, pp. 1–18.

Oaksford, M., Morris, F., Grainger, B. and Williams, J.M.G. (1996) Mood, reasoning and central executive processes. *Journal of Experimental Psychology: Learning, Memory and Cognition*, *22*, 476–492.

Page, M.P.A. and Norris, D. (1998) The primacy model: a new model of immediate serial recall. *Psychological Review*, *105*, 761–781.

Paivio, A. (1986) *Mental Representations: A Dual Coding Approach*. Oxford University Press, Oxford.

Parkinson, L. and Rachman, S. (1981) Intrusive thoughts: the effects of an uncontrived stress. *Advances in Behavioral Research and Therapy*, *3*, 111–118.

Pashler, H. (1990) Do response modality effects support multiprocessor models of divided attention? *Journal of Experimental Psychology: Human Perception and Performance*, *16*, 826–842.

Pavlidis, G.Th. (1985) Erratic eye movements and dyslexia: factors determining their relationship. *Perceptual and Motor Skills*, *60*, 319–322.

Payne, D.G. and Wenger, M.J. (1998) *Cognitive Psychology*. Houghton Mifflin Company, Boston.

Payne, D.G., Peters, L.J., Birkmire, D.P., Bonto, M.A., Anastasi, J.S. and Wenger, M.J. (1994) Effects of speech intelligibility level on concurrent visual task performance. *Human Factors, 36*, 441–475.

Pearson, D.G. and Sahraie, A. (2003) Oculomotor control and the maintenance of spatially and temporally distributed events in visuo–spatial working memory. *Quarterly Journal of Experimental Psychology*.

Pearson, D.G., Logie, R.H. and Gilhooly, K. (1999) Verbal representations and spatial manipulation during mental synthesis. *European Journal of Cognitive Psychology, 11(3)*, 295–314.

Pettito, L.A. and Marentette, P.F. (1991) Babbling in the manual mode: Evidence for the ontogeny of language. *Science, 251*, 1493–1496.

Petty, R. and Cacioppo, J.T. (1986) The Elaboration likelihood model of persuasion. In L. Berkowitz (Ed.), *Advances in Experimental Social Psychology, 19*, 123–205.

Pinker, S. (1994) *The Language Instinct*. William Morrow, New York.

Posner, M.I. (1980) Orienting of attention. *Quarterly Journal of Experimental Psychology, 32*, 3–25.

Premack, D. (1971) Language in chimpanzee? *Science, 172*, 808–822.

Pylyshyn, Z.W. (1973) What the mind's eye tells the mind's brain: A critique of mental imagery. *Psychological Bulletin, 80*, 1–24.

Quinn, J.G. and McConnell, J. (1996) Irrelevant pictures in visual working memory. *Quarterly Journal of Experimental Psychology, 49A*, 200–215.

Ramachandran, V.S. (1988) Perception of shape from shading. *Nature, 331*, 163–166.

Ramachandran, V.S. and Rogers-Ramachandran, D. (1996) Synaesthesia in phantom limbs induced with mirrors. *Proceedings of the Royal Society of London, Series B–Biological Sciences, 263 (1369)*, 377–386

Rayner, K. (1979) Eye guidance in reading: fixation locations within words. *Perception, 8*, 21–30.

Rayner, K. and Pollatsek, A. (1981) Eye movement control during reading: evidence for direct control. *Quarterly Journal of Experimental Psychology, 33A*, 351–373.

Reason, J.T. (1979) Actions not as planned: The price of automatisation. In G. Underwood and R. Stevens (Eds), *Aspects of Consciousness*, Vol. 1: *Psychological Issues*. Academic Press, London.

Reber, A.S. (1967) Implicit learning of artificial grammars. *Journal of Verbal Learning and Verbal Behavior, 6*, 855–863.

Rips, L.J. (1994) *The Psychology of Proof: Deductive Reasoning in Human Thinking*. MIT Press, Cambridge, MA.

Roediger, H.L. and McDermott, K.B. (1995) Creating false memories – remembering words not presented in lists. *Journal of Experimental Psychology: Learning, Memory and Cognition, 21*, 803–814.

Rosch, E. (1973) Natural categories. *Cognitive Psychology, 4*, 328–350.

Rosen, V.M. and Engle, R.W. (1997). The role of working memory capacity in retrieval. *Journal of Experimental Psychology: General, 126(3)*, 211–227.

Roth, E.M. and Shoben, E.J. (1983) The effect of context on the structure of categories. *Cognitive Psychology, 15*, 346–378.

Rovee-Collier, C. and Gerhardstein, P. (1997) The development of infant memory. In N. Cowan (Ed.), *The Development of Memory in Childhood*. Psychology Press, Hove, UK.

Rubin, D.C., Wetzler, S.E. and Nebes, R.D. (1986) Autobiographical memory across the life span. In D.C. Rubin (Ed.), *Autobiographical Memory*. Cambridge University Press, Cambridge.

Rumelhart, D.E. and McClelland, J.L. (1986) *Parallel Distributed Processing: Explorations of the Microstructure of Cognition*. MIT Press, Cambridge, MA.

Sacks, H., Schegloff, E.A. and Jefferson, G.A. (1974) Simplest systematics for the organization of turn taking for conversation. *Language, 50*, 696–735.

Salthouse, T.A. (1991) *Theoretical Perspectives on Cognitive Aging*. Erlbaum, New Jersey.

Sarason, S.B. and Doris, J. (1979) *Educational Handicap, Public Policy, and Social History*. Free Press, New York.

Savage-Rumbaugh, S., Sevcik, R.A. and Hopkins, W.D. (1988) Symbolic crossmodal transfer in two species of chimpanzee. *Child Development, 59*, 617–625.

Savelsberg, G.J.P., Whiting, H.T.A. and Bootsma, R.J. (1991) Grasping tau. *Journal of Experimental Psychology: Human Perception and Performance, 17*, 315–322.

Scarr, S., Pakstis, A.J., Katz, S.H. and Barker, W.B. (1977) The absence of a relationship between degree of white ancestry and intellectual skills within a black population. *Human Genetics, 39*, 69–86.

Schachter, S. and Singer, J. (1962) Cognitive, social and physiological determinants of emotional state. *Psychological Review, 69*, 379–399.

Schank, R.C. (1972) Conceptual dependency: a theory of natural language understanding. *Cognitive Psychology, 3*, 552–631.

Schank, R.C. and Abelson, R.P. (1977) *Scripts, Plans, Goals and Understanding*. Lawrence Erlbaum Associates Inc., Hillsdale, NJ.

Schlaug, G., Jancke, L., Huang, Y.X., Staiger, J.F. and Steinmetz, H. (1995) Increased corpus-callosum size in musicians. *Neuropsychologia, 33*, 1047–1055.

Schneider, W. and Shiffrin, R.M. (1977) Controlled and automatic human information processing: 1. Detection, search, and attention. *Psychological Review, 84*, 1–66.

Scott, S.K., Barnard, P.J. and May, J. (2001) Specifying executive representation and processes in generation tasks. *Quarterly Journal of Experimental Psychology, 54A*, 641–664.

Segal, S.J. and Fusella, V. (1970) Influence of imaged pictures and sounds on detection of visual and auditory signals. *Journal of Experimental Psychology, 83*, 458–464.

Selfridge, O. (1959) Pandemonium: A paradigm for learning. In *The Mechanization of Thought Processes*. H. M. Stationery Office, London.

Senghas, A. and Coppola, M. (2001) Children creating language: How Nicaraguan sign language acquired spatial grammar. *Psychological Science, 12*, 323–328.

Service, E. (1992) Phonology, working memory, and foreign-language learning. *Quarterly Journal of Experimental Psychology, 45A*, 21–50.

Shankweiler, D., Liberman, I.Y., Mark, L.S., Fowler, C.A. and Fischer, F.W. (1979) The speech code and learning to read. *Journal of Experimental Psychology: Human Learning and Memory, 5,* 531–545.

Shepard, R.N. and Metzler, J. (1971) Mental rotation of three-dimensional objects. *Science, 171,* 701–703.

Shin, H.J. and Nosofsky, R.M. (1992) Similarity-scaling studies of dot-pattern classification and recognition. *Journal of Experimental Psychology: General, 121,* 278–304.

Shortliffe, E.H. (1976) *Computer Based Medical Consultations: MYCIN.* Elsevier, New York.

Siegal M., Varley, R. and Want, S.C. (2001) Mind over grammar: Reasoning in aphasia and development. *Trends in Cognitive Sciences, 5,* 296–301.

Simon, H. (1957) *Models of Man: Social and Rational.* Wiley, New York.

Simon, H.A. (1983) *Reason in Human Affairs.* Stanford University Press, Stanford, CA.

Simons, D.J. and Chabris, C.F. (1999) Gorillas in our midst: sustained inattentional blindness for dynamic events. *Perception, 28,* 1059–1074.

Simons, D.J. and Levin, D.T. (1998) Failure to detect changes to people during a real-world interaction. *Psychonomic Bulletin and Review, 5,* 644–649.

Slobin, D.I. (1973) Cognitive pre-requisites for the acquisition of grammar. In C.A. Ferguson and D.I. Slobin (Eds), *Studies of Child Language Development.* Holt, Rinehart & Winston, New York, pp. 175–208.

Smith, E.E., Shoben, E.J. and Rips, L.J. (1974) Structure and process in semantic memory. *Psychological Review, 81,* 214–241.

Smith, S. (1995) Getting into and out of mental ruts. In R.J. Sternberg and J.E. Davidson (Eds), *The Nature of Insight.* Cambridge University Press, Cambridge, pp. 229–251.

Smyth, M.M., Pearson, N.A. and Pendleton, L.R. (1988) Movement and working memory: Patterns and positions in space. *Quarterly Journal of Experimental Psychology, 40(A),* 497–514.

Snowling, M. (1987) *Dyslexia: A Cognitive Developmental Perspective.* Blackwell, Oxford.

Spelke, E.S., Hirst, W.C. and Neisser, U. (1976) Skills of divided attention. *Cognition, 4,* 215–330.

Sperling, G. (1960) The information available in brief visual presentations. *Psychological Monographs, 74,* 1–29.

Standing, L.G., Conezio, J. and Haber, N. (1970) Perception and memory for pictures: Single-trial learning of 2500 visual stimuli. *Psychonomic Science, 19,* 73–74.

Stefflre, V., Castillo Vales, V. and Morley, L. (1966) Language and cognition in Yucatan: a cross-cultural replication. *Journal of Personality and Social Psychology, 4,* 112–115.

Stein, J.F. and Fowler, M.S. (1993) Unstable binocular control in dyslexic children. *Journal of Research in Reading, 16,* 30–45.

Strayer, D.L. and Johnston, W.A. (2001) Driven to distraction: Dual-task studies of simulated driving and conversing on a cellular telephone. *Psychological Science, 12,* 462–466.

Stroop, J.R. (1935) Studies of interference in serial verbal reactions. *Journal of Experimental Psychology, 18,* 643–662.

Tallal, P., Miller, S. and Fitch, R.H. (1993) Neurobiological basis of speech: A case for the pre-eminence of temporal processing. *Annals of the New York Academy of Sciences, 682,* 27–47.

Terrace, H.S. (1979) How Nim Chimpsky changed my mind. *Psychology Today, 3,* 65–76.

Thomas, J.C., Jr. (1974) An analysis of behavior in the hobbits–orcs problem. *Cognitive Psychology, 6,* 257–269.

Thompson, P. (1980) Margaret Thatcher – a new illusion. *Perception, 9,* 483–484.

Thurstone, L.L. (1938) *Primary Mental Abilities.* University of Chicago Press, Chicago.

Tipper, S.P. (1985) The negative priming effect: Inhibitory priming by ignored objects. *Quarterly Journal of Experimental Psychology, 37A,* 571–590.

Tipper, S.P. and Behrmann, M. (1996) Object-centered not scene-based visual neglect. *Journal of Experimental Psychology: Human Perception and Performance, 22,* 1261–1278.

Treisman, A.M. (1964) Verbal cues, language, and meaning in selective attention. *American Journal of Psychology, 77,* 206–219.

Treisman, A.M. (1988) Features and objects: The fourteenth Bartlett memorial lecture. *Quarterly Journal of Experimental Psychology, 40A,* 201–237.

Tulving, E. (1972) Episodic and semantic memory. In E. Tulving and W. Donaldson (Eds), *Organisation of Memory.* Academic Press, London.

Tversky, A. and Kahneman, D. (1974) Judgement under uncertainty: heuristics and biases. *Science, 211,* 1124–1131.

Tversky, A. and Kahneman, D. (1981) The framing of decisions and the psychology of choice. *Science, 211,* 453–458.

Underwood, G. (1974) Moray vs. the rest: The effects of extended shadowing practice. *Quarterly Journal of Experimental Psychology, 26,* 368–372.

Von Wright, J.M., Anderson, K. and Stenman, U. (1975) Generalisation of conditioned G.S.R.s in dichotic listening. In P.M.A. Rabbitt and S. Dornic (Eds), *Attention and Performance V.* Academic Press, London.

Wade, K.A., Garry, M., Read, J.D. and Lindsay, D.S. (2002) A picture is worth a thousand lies: Using false photographs to create false childhood memories. *Psychonomic Bulletin and Review, 9,* 597–603.

Wallas, G. (1926) *The Art of Thought.* Jonathan Cape, London.

Warren, R.M. and Warren, R.P. (1970) Auditory illusions and confusions. *Scientific American, 223,* 30–36.

Warrington, E.K. and Shallice, T. (1969) The selective impairment of auditory verbal short-term memory. *Brain, 92,* 885–896.

Warrington, E.K. and Shallice, T. (1984) Category-specific semantic impairments. *Brain, 107,* 829–853.

Wason, P.C. (1960) On the failure to eliminate hypotheses in a conceptual task. *Quarterly Journal of Experimental Psychology, 12,* 129–140.

Wason, P.C. (1966) Reasoning. In B. Foss (Ed.), *New Horizons in Psychology.* Penguin, Harmondsworth, pp. 135–151.

Watts, F.N., Trezise, J. and Sharrock, R. (1986) Detail and elaboration in phobic imagery. *Behavioural Psychotherapy, 14,* 115–123.

Wegner, D.M. and Wheatley, T. (1999) Apparent mental causation – Sources of the experience of will. *American Psychologist, 54,* 480–492.

Weisberg, R.W. (1995) Prolegomena to theories of insight in problem solving: A taxonomy of problems. In R.J. Sternberg and J.E. Davidson (Eds), *The Nature of Insight*. Cambridge University Press, Cambridge, pp. 157–196.

Weiskrantz, L. (1986) *Blindsight: A Case Study and its Implications*. Oxford University Press, Oxford.

Welford, A.T. (1952) The psychological refractory period and the timing of high speed performance. *British Journal of Psychology, 43*, 2–19.

Werker, J.F. (1995) Exploring developmental changes in cross-language speech perception. In Gleitman, L.R. and Liberman, M. (Eds), *An Invitation to Cognitive Science*, Vol. 1. MIT Press, Cambridge, MA, pp. 87–106.

Wilkins, A. (1995) *Visual Stress*. Oxford, Oxford Science Publications.

Wilkins, M.C. (1928) The effect of changed material on the ability to do formal syllogistic reasoning. *Archives of Psychology, 16 (whole number 102)*.

Winograd, T. (1972) *Understanding Natural Language*. Academic Press, New York.

Woodworth, R.S. and Sells, S.B. (1935) An atmosphere effect in formal syllogistic reasoning. *Journal of Experimental Psychology, 18*, 451–460.

Yerkes, R.M. and Dodson, J.D. (1908) The relation of strength of stimulus to rapidity of habit formation. *Journal of Comparative Neurology and Psychology, 18*, 459–482.

Young, A.W., Hellawell, D.J. and de Haan, E.H.F. (1988) Cross-domain semantic priming in normal subjects and a prosopagnosic patient. *Quarterly Journal of Experimental Psychology, 40A*, 561–580.

Young, A.W., Hellawell, D. and Hay, D.C. (1987) Configurational information in face perception. *Perception, 16*, 747–759.

Zajonc, R.B. (1980) Feeling and thinking: preferences need no inferences. *American Psychologist, 35*, 151–175.

FURTHER READING

Section A – The nature of cognitive psychology

Eysenck, M.W. and Keane, M.T. (2000) *Cognitive Psychology: A Student's Handbook*, 4th Edn. Psychology Press, Hove.

Payne, D.G. and Wenger, M.J. (1998) *Cognitive Psychology*. Houghton Mifflin Company, Boston.

Stanovich, K.E. (1998) *How To Think Straight About Psychology*. Longman, New York.

Section B – Perception

Bruce, V., Green, P.R. and Georgeson, M.A. (2003) *Visual Perception: Physiology, Psychology and Ecology*, 3rd Edn. Psychology Press, Hove.

Gordon, I.E. (1997) *Theories of Visual Perception*, 2nd Edn. Wiley, Chichester.

Moore, B.C.J. (2003) *An Introduction to the Psychology of Hearing*, 5th Edn. Academic Press, London.

Roth, I. and Bruce, V. (1995) *Perception and Representation: Current Issues*, 2nd Edn. Open University Press, London.

Section C – Attention

Styles, E. (1997) *The Psychology of Attentional Behaviour*. Psychology Press, Hove.

Section D – Memory

Andrade, J. (Ed.) (2001) An introduction to working memory. In: *Working Memory in Perspective*. Psychology Press, Hove, UK, pp. 3–30.

Baddeley, A.D. (1997) *Human Memory: Theory and Practice*, Revised Edn. Psychology Press, Hove.

Richardson, J.T.E. (1996) *Working Memory and Human Cognition*. Oxford University Press, Oxford.

Section E – Mental representation

Kosslyn, S.M. (1994) *Image and Brain*. MIT Press, Cambridge, MA.

Gurney, K. (1997) *An Introduction to Neural Networks*. UCL Press, London.

Richardson, J.T.E. (1999) *Imagery*. Psychology Press, Hove.

Section F – Problem solving

Gilhooly, K.J. (1996) *Thinking: Directed, Undirected and Creative*, 3rd Edn. Academic Press, London.

Sternberg, R.J. (1988) *The Nature of Creativity: Contemporary Psychological Perspectives*. Cambridge University Press, Cambridge.

Section G – Reasoning and decision making

Garnham, A. and Oakhill, J. (1994) *Thinking and Reasoning*. Blackwells, Oxford.

Johnson-Laird, P.N. (1983) *Mental Models*. Cambridge University Press, Cambridge.

Tversky, A. and Kahneman, D. (1974) Judgement under uncertainty: heuristics and biases. *Science, 211*, 1124–1131.

Section H – Language

Clark, H.H. and Clark, E.V. (1977) *Psychology and Language: An Introduction to Psycholinguistics*. Harcourt Brace Jovanovich, New York.

Karmiloff, K. and Karmiloff-Smith, A. (2001) *Pathways to Language: From Fetus to Adolescent (The Developing Child)*. Harvard University Press, Boston.

Pinker, S. (1994) *The Language Instinct*. William Morrow, New York.

Section I – Comprehension

Garnham, A. (1985) *Psycholinguistics*. Methuen, London.

Rayner, K. and Pollatsek, I. (1989) *The Psychology of Reading*. NJ, Prentice Hall.

Section J – Intelligence

Cooper, C. (1998) *Individual Differences*. Arnold, London.

Howe, M. (1997) *IQ In Question: The Truth About Intelligence*. Sage, London.

Section K – Consciousness

Baars, B. (1988) *A Cognitive Theory of Consciousness*. Cambridge University Press, Cambridge.

Blackmore, S. (2001) State of the art: Consciousness. *The Psychologist, 14*, 522–525.

Blackmore, S. (2003) Consciousness: An introduction. Hodder and Stoughton, London.

Cleeremans, A., Destrebecqz, A. and Boyer, M. (1998) Implicit learning: news from the front. *Trends in Cognitive Sciences, 2*, 406–416.

Young, A.W. and Block, N. (1996) Consciousness. In V. Bruce (Ed.), *Unsolved Mysteries of the Mind*. Erlbaum, Hove.

Section L – Emotion and mood

Parkinson, B., Totterdell, P., Briner, R.B. and Reynolds, S. (1996) *Changing Moods: The Psychology of Mood and Mood Regulation*. Longman, London.

Strongman, K. (1996) *The Psychology of Emotion*, 4th Edn. Wiley, Chichester.

Williams, J.M.G., Watts, F., MacLeod, C. and Mathews, A. (1997) *Cognitive Psychology and Emotional Disorders*, 2nd Edn. Wiley, Chichester.

INDEX